MORE PEOPLE AND PLACES IN IRISH SCIENCE AND TECHNOLOGY

MORE PEOPLE AND PLACES
IN
IRISH SCIENCE AND TECHNOLOGY

Edited by
**Charles Mollan, William Davis
and Brendan Finucane**

**ROYAL IRISH ACADEMY
1990**

© The Royal Irish Academy 1990
ISBN Hardback 0 901714 82 8
Softback   0 901714 83 6

Published by the Royal Irish Academy for the Royal Irish Academy and EOLAS

Cover design by Alan Corsini

Typeset by Phototype-Set Ltd.
Printed in the Republic of Ireland by Criterion Press

PREFACES

This second volume of pen-portraits of men and women and of associated locations of note in Irish science and technology is indeed welcome. It represents another step towards remedying the parlous condition of the history of science in Ireland which prompted the Academy to establish its National Committee for the History and Philosophy of Science in 1980. This publication will be of value in many ways: by informing the general reader of this important aspect of Irish life; by interesting practising scientists and technologists in the history of their individual disciplines; and, hopefully, by encouraging scholars to undertake full length studies of some of the personalities included.

J.C.I. Dooge
President
Royal Irish Academy

Scientific textbooks and histories of science have tended to overlook the contribution that Irish men and women have made to the development of science. People may be surprised to learn that in Ireland in the seventeenth and eighteenth centuries there was an enormous amount of interest in scientific and technical matters. We had renowned mathematicians, astronomers, physicists, chemists, botanists, technologists and instrument makers, many of whom had a seminal influence in their fields of work. This publication, the second in the series, brings together pen-pictures of Irish people and places who have contributed to the advancement of scientific knowledge.

G.T. Wrixon
Chairman
EOLAS — The Irish Science
and Technology Agency

CONTENTS

CONTENTS IN ALPHABETICAL ORDER

INTRODUCTION

The reception of *Some people and places in Irish science and technology* (which for convenience we shall henceforth call Volume 1) was most gratifying. Although we were fully aware of its imperfections and limitations, it was generously received by most critics. It also sold, and was soon out of print! Indeed the hard-back version of the book, (of which only 250 copies were produced), has become something of a collectors' item, and is already commanding several multiples of its original price on the rare occasions when it appears on the second-hand market. We hope, one day, it can be reprinted, ideally in a new and revised edition, but our preference in the short term is to provide more original material. *More people and places in Irish science and technology* — alias Volume 2 — is the present step in this process.

If Volume 2 is also well received, we shall be emboldened to embark on Volume 3, nameless as yet, which will contain similar brief details of those people who were Irish by birth, but whose major contributions to the development of science and technology were made outside our shores. Among the best known of these are Robert Boyle (1627–91), George Stokes (1819–1902), John Tyndall (1820–93) and William Thomson, Lord Kelvin (1824–1907). As in Volumes 1 and 2, though, we shall not confine ourselves to the giants, but will focus also on lesser, though still important, contributors to the growth of knowledge. Such people are cumulatively, rather than individually, as important as the geniuses.

The most frequent criticism of Volume 1 was, indeed, that some important people, such as Robert Kane (1809–90) and James MacCullagh (1809–47) were excluded, while other less important people were included. The same criticism can be levelled at Volume 2, but it is, in fact, our wish that each of the volumes should represent a range of ability. This may, in turn, require a Volume 4, but enough is enough for the moment.

Meanwhile it is good to note, in the course of the 1980s decade, the growing appreciation of Ireland's scientists at home. Volume 1 has certainly played its part in this. So too has the increasing number of academic books on Irish scientists and science. There are some excellent single-author biographies, like Hankins on Hamilton (1), MacHale on Boole (2) and O'Brien on Corrigan (3). There are collections of essays, for example on Griffith (4) and Mallet (5). There are more general works, like Wayman's history of Dunsink Observatory (6) and Burnett and Morrison-Low's study of the Irish instrument making trade (7). Aer Lingus, which has done so much for young Irish scientists and for the popularisation of science, put on a fine display relating to Dublin-born scientists of the past at its Young Scientists Exhibition during Dublin's Millennium Year 1988, and published a simple booklet to commemorate the exhibition (8).

We look forward to more of the same in the 1990s and to the needed works of synthesis, which will seek to integrate our science and scientists into the mainstreams of both Irish history and international science history. Then someone can popularise the results to complement and enhance the value of our continuing People and Places series.

We should like to record our appreciation to EOLAS for giving support to the efforts of our contributors and ourselves. It is very satisfying to know that, while the major role of EOLAS is in the development of science and technology in support of Irish industrial development, those who make the decisions there are willing to sponsor our efforts in the promotion of scientific awareness. We hope that the 'value added' will be the encouragement of more young Irish people to become involved in science and technology to the commercial and academic benefit of the country, and in the realisation that they are continuing a long an honourable Irish tradition.

Our particular thanks are, of course, due to all our patient contributors who have borne with us over the long gestation period of this volume.

We record our appreciation to the Royal Irish Academy's National Committee for the History and Philosophy of Science, which supported the idea that this second volume should be published and which drew up the list of contents, and to the Royal Irish Academy itself for accepting the recommendation and agreeing to publish the results.

Finally we should like to acknowledge the assistance of Barbara Young and Natasha Weyer-Brown, of the staff of the Royal Irish Academy, and to thank them for treating us so well during the processes of publication.

(1) Thomas Hankins, *Sir William Rowan Hamilton* (Baltimore and London, 1980).
(2) Desmond MacHale, *George Boole — his life and work* (Dublin, 1985).
(3) Eoin O'Brien, *Conscience and conflict — a biography of Sir Dominic Corrigan, 1802–1880* (Dublin, 1983).
(4) Gordon Herries Davies and Charles Mollan (eds), *Richard Griffith 1784–1878* (Dublin, 1980).
(5) Ronald Cox (ed.), *Robert Mallet, F.R.S., 1810–1881* (Dublin, 1982).
(6) Patrick Wayman, *Dunsink Observatory 1785–1985* (Dublin, 1987).
(7) J.E. Burnett and A.D. Morrison-Low, *'Vulgar and mechanick' — the scientific instrument trade in Ireland 1650–1921* (Dublin, 1989).
(8) (Charles Mollan), *Famous Dublin born scientists — Aer Lingus Young Scientists Exhibition '88* (Dublin, 1988).

Charles Mollan
Royal Dublin Society
November 1989

Statue in Armagh Cathedral.

Born: Dublin, 14 April 1661

Died: 19 October 1733

Family: Youngest son of Samuel Molyneux, master gunner for Ireland, and his wife Margaret Dowdall
    Married: Catherine Howard of Shelton, Co. Wicklow, 1693
      Children: several

Distinctions:
Fellow of the Royal Society 1686
President, King's and Queen's College of
    Physicians of Ireland 1702, 1709, 1713, 1720
Professor of medicine, Dublin University,
    1711–33
State physician 1715–30
Physician-general of the army 1718–33

Addresses:
1690–1711   Thomas Court, Dublin
1711–1733   Peter Street, Dublin

The scion of a family that held public office for generations in Ireland, Thomas Molyneux attended Dr Henry Rider's school before entering Trinity College Dublin in 1676. Having graduated M.A. and M.B. in 1683, he went abroad to complete his education. In London, where he arrived about 15 May 1683, he took lodgings 'at the sign of the Flower-de-luce, over against St Dunstan's Church in Fleet Street'. The young doctor visited Gresham College on a day when the Royal Society was meeting and he was permitted to sit in and listen to the discourse. 'At this meeting', he informed his brother, 'I had the opportunity of seeing several noted men as Mr Evelyn, Mr Hooke, Mr Isaac Newton...' While in London he also met Robert Boyle, Sir William Petty and John Flamsteed, who was to be the first astronomer royal.

Molyneux visited Windsor and Eton College where he found the rooms were kept 'very nastily stinking when you come in to them'. He also went to Oxford and Cambridge and in July moved to Leyden, where he met John Locke with whom he later corresponded. In August 1684 his first scientific article, which discussed 'the dissolution and swimming of heavy bodies in Menstruums far lighter than themselves', was published in *Nouvelles de la République des Lettres* and, in December 1684, he sent his first contribution (an account of a 'prodigious *os frontis*') to the *Philosophical Transactions*. He spent some time in Paris in 1685 and was elected a fellow of the Royal Society in the following year.

On his return to Dublin in 1687, Thomas Molyneux proceeded to his M.D. and set up in practice at Thomas Court but, because of the unsettled political situation after the accession of James II, he and many Protestants left Ireland temporarily. For a time he practised in Chester but, after the victory of William of

Orange at the Boyne, he returned to the Irish capital and before long had attained remarkable professional success. He has, indeed, been called the 'father of Irish medicine'.

Thomas and Catherine Molyneux had a large family. They also took charge of a nephew Samuel (a future astronomer), when William Molyneux died in 1698. The latter ('the single most influential figure in Irish science' in his day, according to Alan Gabbey) had held office as secretary and treasurer of the Dublin Philosophical Society. He was the first to demonstrate microscopically the circulation of the blood in reptiles. He represented Dublin University in the Irish parliament of 1692.

Thomas Molyneux was also a member of the Dublin Philosophical Society, which wished to establish experimental science on a proper footing in Ireland. Like his brother, he contributed through moral support, enthusiasm and example rather than by major discoveries. Being an antiquary, geologist and physician, Thomas Molyneux's contributions to the society were correspondingly varied and included an over-credulous account of an operation to remove an ivory bodkin from the bladder of an ingenuous young woman who said she had accidentally swallowed it, a detailed account of the Giant's Causeway, and a paper on 'the late general coughs and colds in Ireland'. Some of his contributions were republished in the *Philosophical Transactions.* He also wrote an account of coal-mining in Ireland.

His acquaintances included Roderick O'Flaherty (1629–1718), and a picture from the early eighteenth century worth keeping in mind is that of Molyneux cantering along a road in Connemara on 21 April 1709 to visit the old historian, at the mercy of an icy wind off the Atlantic which cut through his rain-sodden greatcoat. 'I never saw so strangely stony and wild a country', he reflected later. He and O'Flaherty shared a common interest in scholarship and aptly represented their Anglo-Irish and Irish heritages, one belonging to a privileged, the other to an oppressed class. Both, of course, would have owned to a sense of nationalism. Molyneux's brother, William, had had his book *The case of Ireland stated* burned by the common hangman; O'Flaherty was the author of *A chorographical description of West or h-Iar Connaught.*

Thomas Molyneux's career was in many ways an ideal one: a classical scholar, a man of scientific temperament, a successful practitioner, he wrote on such diverse topics as the Irish round towers, the Irish elk and the ancient Greek and Roman lyres. He represented Ratoath in the Irish parliament from 1695 to 1699 and was one of the founders of the Royal Dublin Society.

Molyneux held important offices and in 1730 he was created a baronet. He amassed considerable wealth — he boasted that he had spent more than Dr Richard Steevens ever made — and in 1711 built a mansion in Peter Street.

Sir Thomas Molyneux died in 1733. His burial place is disputed — Armagh Catheral (where there is a statue of him by Roubiliac) according to the *Dictionary of national biography* and St Audoen's church according to A.M. Fraser. Lady Molyneux lived on for some years in the Peter Street mansion, which in a later period served as the Molynuex Asylum for Blind Women.

Further reading:
Sir William Wilde, *Dublin University Magazine* **18** (1841), 305–27, 470–89, 604–18, 744–63.
A.M. Fraser, 'The Molyneux family'. *Dublin Historical Record* **16** (1960) , 9–15.
K.T. Hoppen, *The common scientist in the seventeenth century* (London, 1970).
A. Gabbey, 'William Molyneux'. In C. Mollan, W. Davis and B. Finucane (eds), *Some people and places in Irish science and technology* (Dublin, 1985).

J. B. Lyons
Department of the History of Medicine
Royal College of Surgeons in Ireland
Dublin

Bust in Royal College of Surgeons in Ireland.

Born: Dublin, 7 June 1777

Died: Dublin, 10 June 1858

Family: Son of Anne (née Verner) and John Crampton, surgeon-dentist
Married: Selina Hamilton Cannon 12 May 1802
Children: two sons and four daughters
The elder son, John Fiennes, became British ambassador to Russia

Distinctions:
M.D. Glasgow 1800
Member of the Royal College of Surgeons in Ireland 1801
President, Royal College of Surgeons 1811, 1820, 1844 and 1855
Surgeon-general 1813
Fellow of the Royal Society
A founder and later president of the Royal Zoological Society
Created a baronet 1839

Addresses:
1777–   16 William Street, Dublin
1813–   24 Dawson Street
About 1813 14 Merrion Square North
            Lough Bray House, Co. Wicklow

'Who is he?': Sir Philip Crampton's memorial fountain bust in Brunswick Street stirred Leopold Bloom's curiosity. The answer (not supplied in *Ulysses*) is that Crampton was a leading Dublin surgeon. The memorial was removed some years ago and today the street where it stood is called Pearse Street.

At fourteen, Philip Crampton was apprenticed to a surgeon, Solomon Richards of York Street, in due course studying at the College of Surgeons (then in Mercer Street) and the Meath Hospital. On 25 September 1798 he took the 'letters testimonial' of the College having held a brief army attachment under the command of Sir John Moore.

Crampton was musical. He liked to play duets with a friend, Theobald Wolfe Tone, and was so engaged when a message reached the young revolutionary that the authorities had been informed of his membership of the United Irishmen. One of the founders of the College of Surgeons, Surgeon William Dease, who died by his own hand in 1798, was rumoured to have feared arrest for a similar proscribed association. Be that as it may, the tragic cloud's silver lining shone down on the newly-qualified surgeon, who was appointed to the vacancy created by Dease's death. He was to hold the appointment as surgeon to the Meath Hospital for almost sixty years.

He was elected M.R.C.S.I. (equivalent to the present college fellowship) in 1801 and a few months later became a member of the Court of Assistants. In 1802 he married Selina Hamilton Cannon, daughter of an army officer. She was a noted beauty but was destined to sustain extensive burns, which were disfiguring and fatal. The Cramptons' Dawson Street residence was equipped with a dissecting-room and lecture theatre. It stood opposite the Richmond Tavern and the story persists that he earned an over-night success by boldly operating to dislodge a lump of meat from a suffocating waiter's windpipe. It is far more likely that he attained his reputation gradually and less dramatically, for he was described as being 'sagacious in diagnosis, ready in resources, dextrous in the use of instruments and sympathetic in his treatment of patients'. He was also appointed surgeon to the Westmoreland Lock Hospital for Venereal Diseases in 1806.

As a teacher, Crampton made no attempt to conceal from his pupils the realities they faced. 'Now, I am sorry to say', he addressed them at the opening of the academic year, 'that the path of clinical observation is neither short nor pleasant; in truth nothing but a full and entire conviction that it is the only path (not the best), but the only path which leads to professional success, could ever induce the medical student to pursue it.'

The Meath Hospital and County Infirmary, 1822.

Having become surgeon-general in 1813 he attended a *levée* in Dublin Castle wearing an impressive dress uniform. Someone asked who was this striking figure. 'He's the surgeon-general', was the reply. Whereupon a wit said, 'I suppose that's the general of the Lancers?'

A well-read man, Crampton was also devoted to outdoor pursuits, especially fox-hunting. Erinensis, the satiric Irish correspondent of *The Lancet,* inferred that Crampton ('the Nimrod of the County Dublin') refused an academic opportunity in the College of Surgeons, preferring 'the saddle to the professor's chair'. Notwithstanding, he was elected to the presidency of the college four times between 1811 and 1855. He was created a baronet by Queen Victoria in 1839.

As well as the house in Merrion Square, the Cramptons owned a country house near Bray in County Wicklow, from where, even when elderly, after an early morning swim he could ride into the city and amputate a limb before breakfast.

His publications include an essay on entropion (inversion of the eyelid) and a description of an avian muscle, which was named *Musculus cramptonius* in his honour. These and other meritorious articles enhanced his reputation but, sparing of praise, Erinensis places them short of the line of partition between talent and genius. Other judges, nevertheless, deemed his anatomical discovery sufficiently important to warrant his election to fellowship of the Royal Society. *Musculus cramptonius,* a tiny muscle in the eyes of birds, arises from a bony hoop which surrounds the cornea and facilitates fine adjustments of avian vision. He was the first Dublin surgeon to perform lithotrity, a method of removing bladder stones by crushing them with a specially designed instrument. He improved the operation then current for correcting cleft palate.

Sir Philip Crampton died at his Merrion Square residence on 10 June 1858. His directions were obeyed when he was buried in Mount Jerome Cemetery—his remains were encased in Roman cement. A few years later the memorial drinking-fountain referred to in *Ulysses* was erected by his friends and admirers as a symbol of health and usefulness.

Further reading:
Sir Charles Cameron, *History of the Royal College of Surgeons in Ireland* (Dublin, 1886).
Jessie Dobson, *Anatomical eponyms* (2nd edition, Edinburgh, 1962).
Martin Fallon, *The sketches of Erinensis* (London, 1979).

J. B. Lyons
Royal College of Surgeons in Ireland
Dublin

A caricature of Dr Dionysius Lardner about the time he was a professor at University College, London.

Born: Dublin, 1793

Died: Naples, 29 April 1859

Family:
Son of a county Clare solicitor, William O'Brien Lardner
  Married:
1. Cecilia Flood, grand-daughter of the Irish parliamentarian, Henry Flood 1815
2. Mary Heaviside, former wife of Captain Richard Heaviside 1849
  Children: Lardner had three children by his first wife and two daughters by his second. In addition he had a natural son by Anne Marie Boursiquot (née Darley) in 1820. Christened Dionysius Lardner Boursiquot, he was to become very well known as Dion Boucicault, one of the nineteenth century's most successful playrights and author of *The Shaughraun.*

Addresses:
1793–1802 88 Marlborough Street
1803–1814 195 Great Britain Street
1815–1820 12 Russell Street
1820–1822 47 Lower Gardiner Street
1823–1828 29 T.C.D.
1828–1830 3 Jermyn Street, London
After 1830 9 Great Queen Street.

Distinctions:
Member of the Royal Irish Academy 1820
Gold medallist of the Royal Dublin Society 1826
(First) professor of natural philosophy and astronomy, University of London 1828–31
Fellow of the Royal Society
Honorary fellow of the Cambridge Philosophical Society
Honorary fellow of the Statistical Society of Paris
Fellow of the Society for Promoting Useful Arts in Scotland

Son of a solicitor, Lardner started work early in his father's Dublin office. Not liking this, he quickly abandoned the law to enter Trinity College at the age of nineteen in 1812. A brilliant student, who won almost all the available prizes in mathematics and philosophy as an undergraduate, he remained in and about the college for the next fifteen years. During this time he took the B.A., M.A., LL.B., LL.D. degrees and appears to have spent much of this time as a 'grinder' of students. His obvious intention was to become a fellow of the college and it was probably with this in view that he took holy orders, as the election of lay fellows was unusual at that time. Election depended also on vacancies arising from the retirement or death of fellows. He was an energetic author and contributed many articles, mostly on science and education, to the *Edinburgh Review*, the literary and scientific magazine then enjoying tremendous success. (Another equally active contributor was the Edinburgh born Brougham, later to become a leader in educational reform in Britain and lord chancellor of England.) Lardner had some reputation as a mathematician. His textbook on analytic geometry (1823) was recommended to the eighteen-year-old Rowan Hamilton by John Brinkley, astronomer royal of Ireland. Lardner failed to be elected to a fellowship in Trinity College, perhaps because the other fellows felt that the showman side of his personality outweighed the scholarly side. As his later career was to show, he was much concerned with the use and application of science to everyday life—an aspect which was to become an important driving force throughout the coming Victorian reign. This philosophy, born of the industrial revolution, was often in conflict with the more sober and 'schoolman' approach of the older, long established, academic communities. In 1828 Lardner moved to London to become, under the patronage of Lord Brougham—by now a powerful member of the House of Commons—the first professor of natural philosophy and astronomy in the newly founded University (College) of London. The new college was the idea of a group of radicals who felt that the time was ripe for a new style university, directed towards the teaching of useful knowledge, similar in character and style to those of France and, especially, of Germany. The pressure for its foundation came too because of the failure of the existing universities of Oxford and Cambridge to admit students other than those belonging to the established Church of England. Some of the most ardent early supporters of the new London College were Jews, Dissenters and Roman Catholics.

The new University got off to a shaky start. There was much wrangling over financial matters, often with Lardner at the centre. A brilliant public lecturer, he devised demonstration equipment that was both extensive and expensive (some of it was still in use in the physics department a century later). However, his courses were not the success he expected and he resigned after five years to give himself over completely to the publishing enterprise by which he is best remembered, the production of the 133-volume *Cabinet cyclopaedia.*

The frontispiece in a volume of Lardner's Cabinet cyclopaedia.

Lardner early identified the clamour for knowledge of things scientific that built up both within and without the scientific community in the numerically increasing, more educated and more affluent middle-class population during the nineteenth century. Before the advent of specialist journals, the encyclopaedia was a favourite way to put facts and opinions about science before the public. The example of Diderot, the French encyclopaedist, shows how powerful an influence such publications could be. His 28-volume work was thought to be so provocative an influence on society that for some time it was suppressed by the French authorities. Not only did Lardner write many of the volumes of the *Cabinet cyclopaedia* himself, he was able to persuade some of the leading scientists of the time, such as William Herschel, the astronomer who discovered Uranus, and David Brewster, a leading authority on optics, to contribute volumes.

Unlike his Irish contemporaries William Rowan Hamilton and James MacCullagh, Lardner cannot be regarded as making any really startling discoveries as a mathematician and physicist. It is through his publications (the British Library lists over a hundred works), and his lectures before such bodies as the British Association for the Advancement of Science and the Society for the Diffusion of Useful Knowledge, that his contribution as a great populariser of science must be recognised. In doing so he made himself rich.

Further reading:
N. Harte and J. North, *The world of University College, London, 1828–1978* (London, 1978).
H.H. Bellot, *University College, London, 1826–1926* (London, 1929).
R. Fawkes, *Dion Boucicault: a biography* (London, 1979).
D. Diderot, *L'Encyclopédie* (Paris, 1751).
D. Lardner, *The cabinet cyclopaedia* (London, 1830–49).
D. Lardner, *The Museum of Science and Arts* (London, 1856).

William J. Davis
Department of Chemistry
Trinity College
Dublin

Portrrait of James Apjohn (now in Trinity College) showing his wet-bulb hygrometer.

Born: Limerick, 1 September 1796

Died: County Dublin, 2 June 1886

Family: Son of Thomas Apjohn, tax collector
Children: three sons, James Henry, Lloyd, and Richard, who became an engineer in 1872

Distinctions:
Professor of chemistry, Royal Cork Institution 1826–28
Professor of chemistry, Royal College of Surgeons 1828–50
Professor of applied chemistry, Trinity College, Dublin 1841–81
Professor of mineralogy, Trinity College, 1845–81
University professor of chemistry Trinity College, 1850–75
Fellow of the Royal Society 1853
Fellow of the Chemical Society 1861
Vice-president of the Royal Irish Academy
Fellow of the King's and Queen's College of Physicians
Awarded the Cunningham Gold Medal of the Royal Irish
    Academy for essay 'A new method of investigating gaseous
    bodies' 1837

Addresses:
1796–1826 Sunville House, High
    Road, Limerick
1846–1857 Lower Baggot Street,
    Dublin
1858–1886 South Hill, Merrion,
    Blackrock, Co. Dublin

James Apjohn was a competent, well-respected and long-lived chemist, who spent 50 years teaching and conducting research, mainly in physical chemistry, in Dublin. He was educated at Tipperary Grammar School and at Trinity College Dublin, to which he went in 1813 at the age of sixteen. He took his arts degree in 1817, before going on to study medicine. He received the bachelor of medicine degree in 1821 and proceeded to his doctorate sixteen years later. During his medical studies he developed a passion for experimental science and thereafter pursued researches in chemistry, his first paper being published in 1821 and his 48th in 1871 at the age of 75.

In 1824 a number of eminent physicians and surgeons established a new private medical school at Parke Street in Dublin and, in the following year, Apjohn was appointed lecturer in chemistry. Three years later he was offered the post of professor of chemistry in the medical school of the Royal College of Surgeons in Ireland. He quickly established a reputation for teaching and attracted students from all over the British Isles. One such, William Gregory, had studied at Giessen with Justus von Liebig; Gregory was to become author of the first organic chemistry textbook in Britain and Ireland, and went on to become professor of chemistry at Edinburgh. Together Apjohn and Gregory analysed the natural product derived from wood, eblanine. The work eventually appeared as a joint paper in the first issue of the *Transactions of the Royal Irish Academy* in 1841.

Although Apjohn was to remain in his post at the College of Surgeons until 1850, he was additionally appointed to the lectureship in applied chemistry and mineralogy in the engineering school at Trinity College in 1841. On the death in 1850 of Francis Barker, the professor of chemistry at Trinity, Apjohn was appointed to the post; he remained in it until his retirement 25 years later.

Apjohn's research interests can be seen to divide into a number of distinct topics. The specific heat of gases interested him throughout his career: his first paper on the subject was published in 1822, the last in 1853. An associated meteorological subject was of particular fascination: the determination of the dew point by the use of the wet bulb hygrometer. His paper of 1835 to the Royal Irish Academy suggested a formula for calculating the dew point: this became known as Apjohn's formula. More practically, in 1850, Apjohn and three other scientists from Trinity College were charged with the task of visiting twelve coastguard stations to give instruction on how to use the meteorological instruments that were deposited there and to help set up tide gauges.

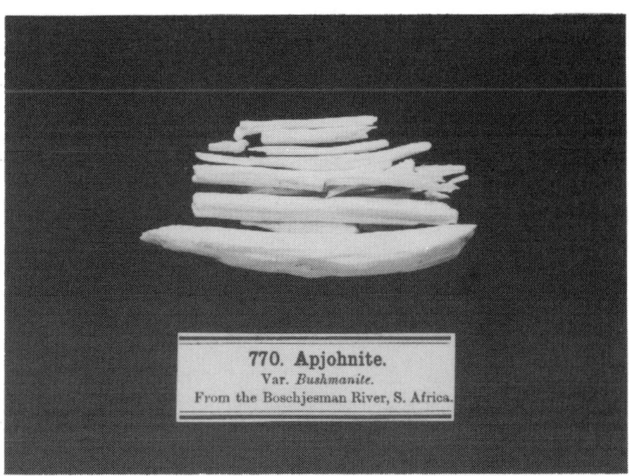

Large crystals of the manganese alum named after Apjohn. (Courtesy of the Trustees of the National Museums of Scotland.)

From the late 1830s Apjohn's interests developed strongly towards mineralogical chemistry. In 1838 he analysed and described a new mineral form Grahamstown, South Africa, which turned out to be a manganese alum. This was subsequently named Apjohnite. Three years later he described a new lead mineral from Kilbricken, Co. Clare, and in 1852 a new yellow–green garnet mineral (Jellettite), which had been found near Zermatt in Switzerland by J. H. Jellett, later provost of Trinity College. Many other analyses of rocks were performed—on a meteorite that had fallen at Adare, Co. Limerick, in 1813, on jade, on hyalite (a glassy silicate with an anomalous polarising effect from Mexico) and on zeolites. Additionally, he investigated some inorganic compounds, especially complexes including potassium iodide. Monographs produced by Apjohn were a descriptive catalogue of the mineral collection at Trinity College (1850), a pamphlet describing the chemical constitution of well water in the college grounds and an inorganic chemistry textbook, *Manual of the metalloids*, first published in 1864, which went into a second edition in 1865.

Apjohn's medical background was not often directly called into use, though he did contribute several articles to the *Cyclopaedia of practical medicine* of 1833–35. (It is said that Charles Dickens founded his account of the death of Krook in *Bleak House* on Apjohn's description of spontaneous combustion.) He also carefully checked the chemical processes described in the first edition of the *British pharmacopoeia* of 1864. After his retirement from the chair of chemistry at Trinity in 1874 he gradually gave up his other professional interests. He died at the advanced age of 90.

Further reading:
Obituary notices in *Journal of the Chemical Society* **51** (1887), 469–70, and *Chemical News* **53** (1886), 296.
E.M. Philbin, 'Chemistry'. In T. Ó Raifeartaigh (ed.), *The Royal Irish Academy: a bicentennial history 1785–1985* (Dublin, 1985), 275–300.

R.G.W. Anderson
National Museums of Scotland
Edinburgh

Thomas Grubb, from a family photograph.

Born: Kilkenny, 1800

Died: Dublin, 1878

Family: Son of William Grubb (d. 1831) and
Eleanor Fayle
   Married: Sarah Palmer 1826
   Children: Four daughters and four sons

Distinctions:
Member of the Royal Irish Academy 1839
Fellow of the Royal Society, London 1864

Address
c. 1850 1 Upper Charlemont Street, Dublin

The family appears to originate in an ancient and noble line of Grubbe in Denmark and North Germany that includes among its members a maternal ancestor of Tycho de Brahe. John Grubb (1620–90) arrived at Waterford from England in 1656 as part of the Cromwellian settlement. His father had been an Anabaptist preacher in Northamptonshire. Some large families are recorded among his descendants, many of whom were members of the Society of Friends. His great-grandson, William Grubb, was cousin to a family of twelve that is still associated with Castle Grace, Tipperary, and was himself one of fourteen children. Marrying twice he had six children, one of whom was Thomas Grubb.

After a start in commercial life, Thomas developed a strong interest in machine tools and, by the 1830s, had founded the Dublin firm of optical engineers and telescope makers that was to last until 1925, when it was merged with C.A. Parsons Ltd of Newcastle-upon-Tyne. Thomas Grubb's youngest son Howard took responsibility for his father's firm in the 1870s, before the age of thirty, and carried on into his eighties. He became Sir Howard Grubb in 1887 (Vol. 1, p.60).

In the years around 1900 there were, in the whole world, only five or six firms capable of making large astronomical telescopes. One of these was that of Grubb in Dublin, later Sir Howard Grubb, Parsons and Company of Newcastle-upon-Tyne. The sound basis on which these achievements rested was the creation of Thomas Grubb, who established a workshop near Charlemont Bridge on the Grand Canal in the 1830s. This district and neighbouring Rathmines was the locality of the Grubbs' work for nearly one hundred years.

In the 1830s Thomas Grubb started on the design and construction of what were then considered large telescopes; they were in fact among the largest of their day. He made instruments for Armagh Observatory, for Mr E.J. Cooper at Markree Castle, Co. Sligo (Vol. 1, p. 28), and for Greenwich Observatory, and in 1852 he made the telescope that eventually became the South 12-inch refractor at Dunsink Observatory.

Thomas Grubb assisted the third earl of Rosse (Vol. 1, p. 85) in the construction of the mirror-cell of his huge six-foot reflector in the 1840s. The two largest instruments which were made in Dublin were the 48-inch Melbourne reflector, completed in 1869, and the 27-inch Vienna refractor, completed in 1880, largely

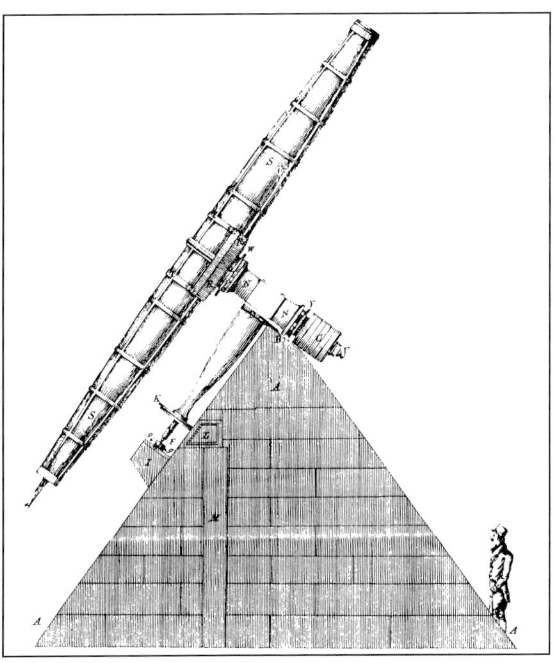

Line drawing of Thomas Grubb's telescope constructed in 1835 for Mr Edward J. Cooper of Markree Castle, Co. Sligo. The lens was made by Cauchoix of Paris with an aperture of 0.34m (13.5 in.) and the focal length was approximately 8.7m.

The 48-inch Melbourne reflector of 1865 made by Thomas Grubb of Dublin.

the work of Howard Grubb. The design of rotating domes was also a Grubb speciality and the firm provided four domes for the new Vienna Observatory at the same time as the telescope was supplied.

In 1853 Thomas Grubb became engineer to the Bank of Ireland. For some fifteen years he had been designing and building machines for printing banknotes to the meticulous standards required, with automatic serial numbers impressed upon them. These machines were designed to carry out copper-plate printing without appreciable alteration in the details after many thousands of impressions. He perfected a means for recutting the copperplates in order to preserve the fidelity of the process. In 1855, he produced argument for preferring his methods to those of the Bank of England, which depended on watermarked paper. His machines worked almost faultlessly into the 1920s. The firm also made machine tools, lenses of all sizes, magnetometers, and at least one microscope of superior design now to be seen in the Oxford Museum of the History of Science.

Thomas Grubb's work was characterised by solid, robust design and careful workmanship. As a consequence his firm had achieved an international reputation by the end of the nineteenth century.

Further reading:
H.C. King, *History of the telescope* (London, 1956).
T.H. Mason, 'Dublin opticians and instrument makers'. *Dublin Historical Record* **11** (1952), 133–49.
G.E. Manville, 'Two fathers and two sons'. *Parsons Journal* (C. A Parsons & Co. Ltd.) **11** (1967), 144–60

Patrick A. Wayman
Dunsink Observatory
Dublin

The third earl of Enniskillen.

Born: Florence Court, 25 January 1807

Died: County Fermanagh, 12 November 1886

Family: The earls of Enniskillen were prominent landowners settled in County Fermanagh since the Plantation of Ulster in 1610. The third earl succeeded to the title in 1840.
   Married:
1. Jane Casamaijor 1844 (died in 1855)
2. Hon. Mary Emma Broderick 1865, daughter of the 6th Viscount Midleton
   Children: There were seven children from his first marriage.

Distinctions:
Fellow of the Geological Society 1828
Fellow of the Royal Society 1829
M.P. for Fermanagh 1831–40; afterwards sat
   in the House of Lords.
Member of the Royal Irish Academy 1846
First president of the Royal Geological Society
   of Ireland 1865
Honorary doctor of civil law Oxford University
Honorary LL.D. universities of Dublin
   and Durham

Little is known of Viscount Cole's boyhood days except that he was educated at Harrow School, but he must have developed an early interest in geology, for, on going to Christ Church, Oxford, in 1826 at the age of 19, he enrolled in the classes of William Buckland, first professor of geology at the university. Buckland, an eccentric but talented teacher, inspired Cole with an enduring passion for the study of fossils. At college Cole met another young aristocrat and fellow student, Sir Philip Egerton, and the pair became life-long friends and later collaborators in collecting fossil fish, for which both became famous.

Cole and Egerton made their first foreign journey in the summer of 1830, when they travelled through Europe on a geological Grand Tour. Thus they met, in Munich Museum, Louis Agassiz, who was to influence profoundly the future course of their geological researches. Agassiz, a brilliant naturalist, was trying to make a complete classification of all the fishes, both fossil and living, and suggested that they could help by collecting all the different kinds of fossil fish in the world. The pair, realising that this was an entirely new field of research, henceforth made the collecting of fossil fish their main scientific pursuit. They returned home later that year, laden with some of the finest fossil fish that could be found in Europe.

Lord Enniskillen was a convivial man who enjoyed good company. In 1828, at the age of 21, he had joined the Geological Society of London, where he met the leading geologists of his day, playing host to them at his London rooms. They knew of his ambition to form a comprehensive collection of fossil fish and frequently helped to obtain for him rare or newly discovered examples. For his part, Lord Enniskillen never let slip any opportunity to secure a new specimen, and he used all his charm and persistence to obtain what he wanted, by purchase, exchange or unashamed begging from friends. By the late 1830s he had many hundreds of these fossils arranged in his private museum at Florence Court — in a pavilion in one of the flanking wings to the main house. He inserted a large lantern window in the roof for

Florence Court, 1870. The Pavilion, left, housed the third earl of Enniskillen's private museum. Photograph: courtesy earl of Erne.

illumination and lined the walls with cases to display his treasures. They eventually numbered nearly 10,000 individual objects, most of them fossil fish. They were the best obtainable, and many were important to science. It remains one of the largest and scientifically most important collections of fossil fish ever assembled.

As well as his parliamentary duties, Lord Enniskillen was prominent in other public affairs, especially in the Orange Order, of which he was an enthusiastic member for over fifty years. This brought him controversy and danger, but his loyalty never wavered and he became first imperial grand master of the order world-wide in 1866.

Lord Enniskillen transferred his collection to the British Museum in London in 1883 (for £3500) where it was arranged in the newly opened Natural History building, alongside the collection of his old friend Sir Philip Egerton, who had died a few years earlier. Lord Enniskillen himself died shortly after at the age of 79 and was buried in the family vault in St Macartan's Cathedral, Enniskillen.

Lord Enniskillen published nothing on his scientific work. His contribution to science was his collection and his generosity in allowing others to use it. His specimens figured—and continue to figure—in the standard works on fossil fish.

Today Florence Court is owned by the National Trust, but there is nothing left there to remind the visitor of the third earl of Enniskillen's scientific life. He deserves to be more widely known as a gracious Irishman, who lived life to the full and whose legacy to geology is safely preserved in the Natural History Museum, London.

Further reading:
K.W. James, *Damned Nonsense!—the geological career of the third earl of Enniskillen* (Belfast, 1986).

Kenneth W. James
Geology Department
Ulster Museum
Belfast

Marble bust by Christopher Moore, R.H.A. (Courtesy of the provost, fellows and scholars of Trinity College Dublin.)

Born: Landahussy, County Tyrone 1809

Died: Dublin, 24 October 1847

Family: Son of James (1777–1857) and Margaret (1784–1839) MacCullagh and the eldest of the eight of their twelve children who survived beyond childhood. The family circumstances were modest but two of his brothers were to follow him to Trinity College. One of these, John (1811–91), was subsequently called to the Bar and appointed a resident magistrate.

MacCullagh never married.

Distinctions:
Fellow of Trinity College Dublin 1832
Member of the Royal Irish Academy 1833
Erasmus Smith's professor of mathematics,
    University of Dublin 1835–43
Cunningham Medal of the Royal Irish Academy 1838
Copley Medal of the Royal Society, London, 1842
Secretary of the Royal Irish Academy 1842–46
Erasmus Smith's professor of natural and experimental
    philosophy, Trinity College Dublin 1843–47
Fellow of the Royal Society, London 1843

Addresses:
1809–1824   Landahussy, Upper Badoney, and Strabane, both in County Tyrone
1826–1847   Trinity College Dublin

MacCullagh's first college rooms were in house number 14; he subsequently moved to number 7 and was living in number 40 when he died.

James MacCullagh was only fifteen when he was admitted to Trinity College Dublin in 1824. It was a good time to study mathematics in Trinity. A major reform of the syllabus had just been implemented, chiefly owing to the efforts of Bartholomew Lloyd, who held the chairs of mathematics and natural philosophy in succession before being elected provost in 1831. There were several able mathematicians among the fellows of the college and the lectures and prescribed texts reflected the new developments in mathematics taking place on the Continent, especially in France. Among his near contemporaries as undergraduates, both a few years senior to MacCullagh and both to become his colleagues as professors in the University of Dublin, were Humphrey Lloyd (Vol.1, p.32), the son of Bartholomew, and William Rowan Hamilton, who was to achieve international renown as a mathematician (Vol.1, p. 36).

MacCullagh's real mathematical talent lay in geometry. He was especially interested in the ellipsoid and other, so-called, surfaces of the second order. As professor of mathematics he lectured on geometry, and exercised a profound influence on a generation of students some of whom, such as George Salmon, were to make their own major contributions in the subject. MacCullagh is, however, better known for his work

Sketches of James MacCullagh made after his death by F.W. Burton. (Courtesy of the National Gallery of Ireland.)

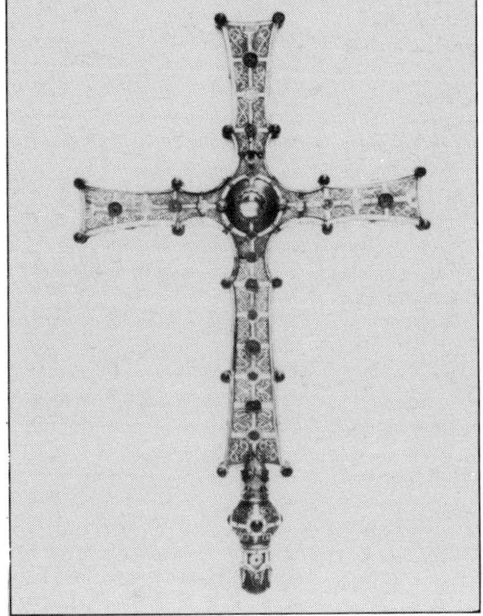

The Cross of Cong, MacCullagh's gift to the nation.

in mathematical physics, particularly in the development of mathematical models for the aether. One of the major scientific quests throughout the nineteenth century was to identify and to describe in mathematical and mechanical terms this medium, to which the name aether had been given, whose existence was supposed to be essential in order for light waves to propagate. It was only in the opening years of this century, following the experimental observations of Michelson and Morley and the radical re-interpretation of basic physical concepts by Einstein in his Special Theory of Relativity, that the aether hypothesis was put aside. MacCullagh's models combined mathematical subtlety, which used his skill as a geometer to the full, with a profound appreciation of the essential physics of light propagation. After his death, when it came to be realised that light was an aspect of the wider and newly discovered phenomenon of electro-magnetic radiation, Fitzgerald showed that McCullagh's aether could equally well describe the more general phenomenon.

MacCullagh played a key role in building up the Academy's collection of Irish antiquities, now housed in the National Museum. Although not a wealthy man, he himself bought the Cross of Cong for 100 guineas to present to the Academy and contributed towards the purchase of other treasures. (This was the start of the collection now in the National Museum.) In August 1847 he stood, unsuccessfully, as parliamentary candidate for one of the Dublin University seats. Later that year, in one of his periodic bouts of depression, he took his own life. An obituary notice in *The Nation* described him as 'a warm and ardent Nationalist'. This is misleading; he was not a Nationalist, but he was a patriotic Irishman devoted to his country and profoundly critical of the lack of national self-respect which he saw about him.

Further reading:
Entries in the *Dictionary of national biography* and the *Dictionary of scientific biography* and obituary notices of the Royal Irish Academy and the Royal Society.
T.D. Spearman, 'James MacCullagh 1809–1847'. In J. Nudds, N. McMillan, D. Weaire and S. McKenna Lawlor (eds), *Science in Ireland 1800–1930* (Dublin, 1988).
B.K.P. Scaife, 'James Mac Cullagh, M.R.I.A., F.R.S., 1809–1847'. *Proceedings of the Royal Irish Academy* (forthcoming).

David Spearman
School of Mathematics
Trinity College
Dublin

> **Born:** Dublin, 24 September 1809
>
> **Died:** Dublin, 16 February 1890
>
> **Family:** Son of John Kane, a member of the United Irishmen, who studied chemistry while a refugee in Paris and on his return to Dublin started a chemical works, making lime, sulphuric acid and bleaching powder.
> Married: Katherine Baily of Newbury, Berkshire 1890
> Children: Four sons, Robert Romney, Henry Coey, Francis Bailey, Roderick, and a daughter, Mrs Valentine Coppinger, Florence

Addresses:

1809–1832	48 Henry Street, Dublin
1832–1845	23 Gloucester Street, North (now lower Sean MacDermott Street)
1846–1856	Gracefield, Booterstown
1857–1873	Wickham, Dundrum
1874–1878	21 Raglan Road
1879–1889	Fortlands, Killiney
1889–1890	2 Wellington Road

Distinctions:
Professor of chemistry, Apothecaries' Hall 1831–45
Member of the Royal Irish Academy 1832
Lecturer in natural philosophy and (later) professor of chemistry at the Royal Dublin Society 1834–47
Director, Museum of Irish Industry 1845
Knighted 1846
First president, Queen's College Cork 1846–73
Fellow of the Royal Society 1849
Honorary LL.D. University of Dublin 1868

Even if it were allowed nowadays, to take seven years to obtain a pass B.A. degree would not be considered an auspicious start to a career in science and education. This is how long Robert John Kane took to obtain his degree in Trinity College, where he became a student in 1828. However, to have published several learned papers on chemistry, to have qualified to practise medicine and won a gold medal in doing so, to have successfully analysed a mineral as the arsenide of manganese and have it called after you (in this case Kaneite), to have become a professor of chemistry, to have been elected a member of the Royal Irish Academy, to have published a textbook on pharmacy, and to have founded a scientific journal that still continues in print today, over the same period of seven years, may perhaps excuse Kane's tardiness in completing his studies as an undergraduate at Trinity College.

Despite his brilliant medical beginnings, Kane's main scientific interests were firmly attached to chemistry. His contributions to chemical knowledge were many. While working in Liebig's laboratory at Giessen in the German province of Hesse in 1836 he isolated acetone, $(CH_3)_2CO$, from wood spirit. On returning to Ireland and while working at the laboratory of the Royal Dublin Society, he changed this simple aliphatic compound into an aromatic hydrocarbon, which he called mesitylene, C_9H_{12}. His method was to distil acetone carefully with sulphuric acid. It is the symmetrical tri-methyl benzene and it was the first cyclic hydrocarbon to be synthesised from a straight chain compound. Kane was an active chemical experimenter and a bold theorist. He was also the first to suggest that alcohol, ether and some esters contained a 'radical', C_2H_5. His researches on the combination of ammonia with metallic salts acquired for him a European reputation. Both the Royal Irish Academy and the Royal Society, London, acknowledged his work on the chemistry of the natural dyestuffs, achil and litmus, which can be extracted from lichens,

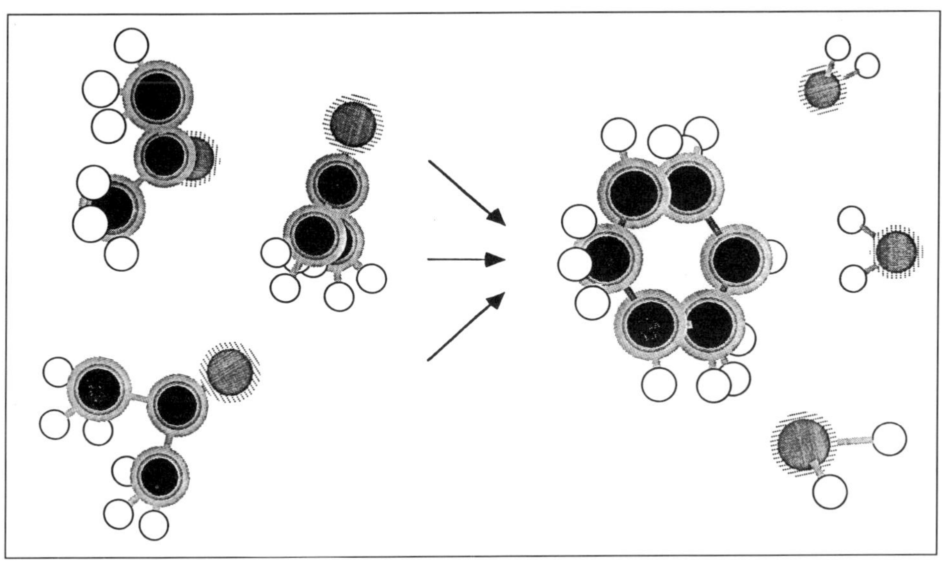

Computer-generated simulation of Kane's synthesis of mesitylene from acetone.

by presenting him with medals. Kane's role as an active chemical researcher ended after 1841 when he published a three-volume textbook on chemistry, which was to become a best-seller. In 1844 he published his brilliant and detailed assessment of Ireland's natural resources and sources of industrial power.

Concurrently with all these activities Sir Robert (he was knighted in 1846) was a very busy administrator. Not only was he advising the government about education, at the national level as well as at university level, but during the Famine years he was giving advice, as one of the eight Irish relief commissioners, on food distribution and later, as a member of the Committee of Health, on how to deal with the outbreaks of typhus fever which were occurring all over the country as a result of the terrible conditions. As director of the Museum of Irish Industry, an institution which was what we now would call a research association, he oversaw a number of investigations, including one into the possibility of growing beet as a source of sugar. The Museum also involved itself in the schemes for the provision of teaching in applied sciences, and some of its staff were incorporated into the Royal College of Science when it was founded in 1867.

After Kane was appointed the first president of the newly formed Queen's College Cork, he continued as director of the Museum of Industry in Dublin for many years. His non-residence in Cork over the 28 years for which he held the latter appointment caused a great deal of controversy. A man of strong opinions, his influence on the moulding of science and education in nineteenth-century Ireland was immense.

Further reading:

T.S. Wheeler and others, *The natural resources of Ireland — a series of discourses delivered before the Royal Dublin Society* (Dublin, 1944).

R. Kane, *The industrial resources of Ireland* (Dublin, 1844).

D. Reilly, *Sir Robert Kane* (Cork, 1942).

D. Reilly, 'Robert John Kane (1809–90): Irish chemist and educator'. *Journal of Chemical Education* **32** (1955), 404.

B. Kelham, 'Robert Kane'. *Studies* **56** (1967), 297.

William J. Davis
Department of Chemistry
Trinity College
Dublin

Pencil sketch c. 1850 by F.W. Burton — courtesy of the
National Gallery of Ireland.

Born: Summerville House, Limerick,
5 February 1811

Died: Torquay, England, 15 May 1866

Family: Son of Joseph Massey and Rebecca
(née Mark) Harvey
  Married: Elizabeth Lecky Phelps in Limerick
on 2 April 1861
  There were no children.

As a child Willliam Harvey often spent his summer holidays at Milltown Malbay on the coast of County Clare where he became fascinated by the sea, by the flotsam it deposited on the beach and by the shells and sea-weeds. Sea-weeds (marine algae) became his lifelong passion.

The Harveys were Quakers, and William, who was the youngest of eleven children, went to the great Quaker schools at Newtown, Co. Waterford, and Ballitore, Co. Kildare, where his love of natural history was encouraged. After he left school William took an interest in non-flowering plants — mosses and liverworts, for example — and this led to his first contacts with Professor William Hooker, who was to become a close friend.

Harvey's mother died in 1831 and after the death of his father in 1834 William sought a government position overseas. Through the good offices of Cecil Spring Rice, M.P. for Limerick, his brother Joseph was appointed colonial treasurer at Cape Town — this was an error, for William should have been appointed. However, William decided to accompany his brother, hoping to spend his time collecting and studying the flora of the Cape of Good Hope, but Joseph Harvey became ill soon after reaching Cape Town and he died two weeks after beginning a return voyage to Europe. William was then appointed in his brother's stead and returned to Cape Town in 1836. He resumed his interrupted botanical studies, joking that he spent more time collecting plants than collecting taxes and thus should be called 'her majesty's pleasurer-general not treasurer-general'.

William Harvey too became ill during his second tour of duty and resigned his position in 1841. However, he had done valuable work on South African botany and encouraged others to collect plants for him; all this was to lead to a series of important publications, including the first three volumes of *Flora Capensis.*

Harvey returned to Ireland and in 1844, after the death of Dr Thomas Coulter, he was appointed curator of the herbarium in Trinity College Dublin; he had no teaching duties, only the care of the collection of dried and labelled plants that comprised the herbarium. He spent many hours sorting, identifying and naming the specimens and publishing the results of his studies. Thousands of specimens from overseas collectors were sent to Trinity College for Harvey to catalogue and study, especially specimens from the Cape of Good Hope, Natal and Transvaal, and also seaweeds.

In 1848 Harvey was elected professor of botany to the Royal Dublin Society; this position obliged him to give public lectures throughout Ireland. Eight years later he was appointed professor of botany in the University of Dublin (Trinity College), and he held both of these professorships until his death. He visited the United States in 1849–50 and undertook a tour to Australia, New Zealand and the South Pacific in 1854–56 to collect seaweeds and other plants.

William Henry Harvey published three important works on the botany of southern Africa, and he also produced three monographs on seaweeds, one on Irish and British species *(Phycologia Britannica)* and the others on species from North America and Australia. He illustrated some of these with his own lithographic drawings for he was a skilful, self-taught draughtsman.

During his career Harvey classified and named hundreds of new species of flowering and non-flowering plants — for example the beautiful Californian tree-poppy, *Romneya coulteri*, after his predecessor Thomas Coulter and the Armagh astronomer Thomas Romney Robinson. He was himself commemorated by a South African genus *Harveya,* a group of parasites related to the European foxgloves. This amused William Harvey, whose voluminous correspondence displays his twinkling sense of humour — frequently he made fun of his pompous fellow botanists.

Principal publications:

1836   *The genera of South African plants*
1844   *The seaside book*
1846–51 *Phycologia Britannica*
1847   *Nereis Australis*
1851–53 *Nereis Boreali-Americana*
1859–63 *Thesaurus Capensis*
1859–65 (with Otto Sonder) *Flora Capensis* (not completed)

Sources and further reading:

S.C. Ducker (ed.), *The contented botanist. Letters of W.H. Harvey about Australia and the Pacific* (Melbourne, 1988).
Lydia Fisher, *Memoir of W.H. Harvey* (London, 1869).
R.L. Praeger, 'William Henry Harvey'. In F.W. Oliver (ed.), *Makers of British botany* (Cambridge, 1916) 20–24.
D.A. Webb, 'W.H. Harvey and the tradition of systematic botany at T.C.D.' *Hermathena* **103** (1966), 32–45.
E.C. Nelson and E.M. McCracken, *The brightest jewel: a history of the National Botanic Gardens, Glasnevin, Dublin* (Kilkenny, 1987).

E. Charles Nelson
National Botanic Gardens
Glasnevin
Dublin

Examining a stereoscopic transparency.

Born: Heaton near Bradford, England
23 July 1813

Died: London 1885

Family: Elder daughter of John Wilmer Field,
a wealthy Yorkshire landowner, and his wife
Anne, née Wharton-Myddelton
    Married: William, third earl of Rosse
    Children: Eleven — four sons (Laurence,
Randal, Clere and Charles) reached adulthood.

Distinctions:
Silver medal of the Photographic Society of
    Ireland 1859
Contributed to the Dublin International
    Exhibition 1865

Following the death of his wife whilst Mary and her sister Delia were very young, Squire Field entrusted their upbringing to Susan Lawson, who recognised Mary's artistic talent and stimulated its development. Evidence of this artistic ability is to be seen in the furniture she designed for Birr Castle and the extensive redesigning of the demesne which she undertook in conjunction with her uncle, Richard Wharton-Myddelton. Two of her most notable achievements are the entrance gates and a drawbridge tower known as the 'keep gate', for both of which she made beautiful scale models. Her interest in heraldry was expressed in the embellishment of her wrought iron gates installed in the 'keep gate'. By the close of 1853, work on the remodelling of the demesne was nearing completion: she had extended the castle to house her large family though sadly this proved to be unnecessary, as only four of her children survived their teens.

At this time, presumably seeking a new challenge, she turned to photography. It would appear that she first experimented with the recently introduced 'wet collodion' process and with the 'daguerreotype' apparatus purchased by her husband in 1842. These first photographs, which were of the great telescope, were sent by her husband to W.H. Fox Talbot, who responded in most complimentary terms requesting permission to have them framed and exhibited at what was to be the first exhibition of the London Photographic Society, later to become the Royal Photographic Society.

The countess created a fine darkroom in the castle and gradually acquired a wide range of cameras, including an early single lens stereo camera by Knight of London which was used to make many of the stereoscopic views in the collection. She also used a superb 12" x 15" 'folding' double box camera, with which she took pictures by both the collodion and the 'waxed paper' processes.

During the following years, Mary Rosse exhibited her work in Ireland and England and became a recognised exponent of the 'waxed paper' process. Her pictures received favourable comment in the photographic journals and the Photographic Society of Ireland awarded her a silver medal. In 1865, the jury of the Dublin International Exhibition made special mention of her photographic exhibit. She became

Richard Wharton-Myddelton, Mary Rosse's uncle. He was deeply involved in the design of the improvements to the castle and Demesne undertaken by the Countess. This is one of her most powerful portraits.

The six-foot telescope from the south-west. A group of family members is composed within this picture. The man wearing a top hat is the third earl , whilst Charles Parsons is seated on the steps. Circa 1857-58.

well known as an active member of the Amateur Photography Association, through which organisation her work, including general views and detail pictures of the six-foot telescope, was widely distributed.

Although the countess mastered the complex chemistry of several photographic processes and perfected a form of the waxed paper process, her scientific and technical accomplishment was merely a means to an end. Her greatest achievement was artistic. Her superb compositions, whether individual portraits, groups or landscapes, are of the highest order. Her ability to infuse an atmosphere of spontaneity whilst making exposures of around twenty seconds is quite remarkable. It is unfortunate that only one of her paper negatives is known, for this process allowed great flexibility to the photographer as the material could be prepared in advance, used and processed when convenient, being suited thereby to use when out of reach of a darkroom. It is tempting to speculate on the content and quality of these negatives, especially as records show that she took pictures in Spain whilst on a cruise in the earl's yacht 'Titania'.

It would seem that she took no further photographs after the death of her husband in 1867, and, as her relationship with the new countess became increasingly strained, she moved out of the castle and, following a brief sojourn in Dublin, she moved in 1870 to the house in Connaught Place, London, where she and her husband had entertained most of the leading scientists of the era, and hosted the functions at which Talbot's 'photoglyphic engravings' and the world's first news photograph were first shown. She lived at this address until her death in 1885. She was buried in Birr beside her husband. Her four sons commissioned Robert Kemp to design a memorial window which is still to be seen in the parish church at Heaton.

Further reading:
David H. Davison, *Impressions of an Irish countess* (Birr, 1989).

David H. Davison
Dundrum
Dublin

Born: Beech Hill, Co. Armagh, 15 March 1815

Died: London, 26 February 1902

Family: The youngest of nine children of Thomas Simpson
 Married: Mary Martin of Loughorne, Co. Down, a sister of John Martin, a Young Irelander
 Children: Two daughters

Distinctions:
Fellow of the Chemical Society London 1857,
 Vice-President 1872–74
Fellow of the Royal Society 1862
Honorary fellow of the King's and Queen's College
 of Physicians of Ireland 1865
Professor of chemistry, Queen's College Cork
 1872–92
President, chemical section of the British
 Association 1878
Honorary LL.D. University of Dublin 1878
Honorary D.Sc. Queen's University 1882

Addresses:
1815–1845 Beech Hill, Co. Armagh
1845–1857 11 Wellington Road, Dublin
1860–1870 33 Wellington Road, Dublin
1870–1892 11 Dyke Parade, Cork
1893–1902 7 Darnley Road, Holland Park, London

Maxwell Simpson's long life spanned a major portion of the period in which the first developments of modern chemistry were being made. In the year of his birth Faraday was 24 and was working on the chemistry of simple hydrocarbons; his electrical researches were still about ten years away. In that same year Humphry Davy invented the miner's safety lamp, Dalton was 49, Berzelius was 36, Liebig was a boy of 12, Kekulé was not yet born. When Simpson died in 1902, Planck's quantum theory was just two years old and the scene was set for the great leap in the understanding of physical science.

Simpson received his early education in Newry at the same school attended by John Martin and John Mitchel, both of whom were to make a considerable impact in Irish politics as 'Young Irelanders'. In 1832 Simpson went on to study medicine at Trinity College where Martin and Mitchel were also students. Simpson did not, at that time, complete his medical course, but left Trinity in 1837 with the degree of Bachelor of Arts.

On a visit to Paris a few years later he attended a lecture on chemistry given by Jean Baptiste André Dumas. This seems to have made such an impression on Simpson that he decided to take up the study of chemistry. For two years he attended the lectures at London University given by Thomas Graham, best known for his famous law on the diffusion of gases. He also worked in Graham's laboratory. In 1845 Simpson came back to his friends of student days in Dublin, and married the sister of 'Honest John' Martin. He seems to have maintained his friendship with the United Ireland leaders throughout his scientific career in Dublin and later in Cork: shortly before he was transported to Van Diemen's land (now Tasmania) Martin records a visit from Simpson.

Justin Von Liebig's famous laboratory at Giessen 1842.

After his marriage Simpson resumed his medical studies so that he could become a lecturer in chemistry in the Park Street Medical School in Dublin. He qualified in 1847. In 1849 he moved to a similar post in the Peter Street or 'Original' School of Medicine. While in this position he was granted three years' leave of absence, during which he studied in Germany under Adolph Kolbe in Marburg and Robert Bunsen in Heidelberg.

Simpson then returned to Dublin and continued to lecture till 1857, when he resigned and went to Paris. Here he assisted Wurtz, who had discovered methylamine and ethylamine in 1849. These men with whom he collaborated were some of the foremost organic chemists in Europe. After two years with Wurtz he returned to Dublin, where he set up his own laboratory in a back kitchen of his house in Wellington Road. It was here that Simpson carried out many of the researches that have made him a significant figure in the history of Irish chemistry. During the eight years which Simpson spent at work in his home-based laboratory not a season passed without his publishing one or two papers of first-class scientific importance. He was expert both in organic synthesis and in the elemental analysis which was needed to get information about the mechanisms of reactions. He was the first chemist to synthesise succinic acid, $C_2H_4(COOH)_2$, until then only available from natural sources, from ethylene cyanide. He was also first to point out that the acidity of organic acids arises from the presence in the molecule of what we would now call the functional carboxylic group, –COOH, but which he called the 'semi-molecule of oxatyl'. As well as working on di-and tri-basic acids he also synthesised many new halogen-containing organic compounds. His publications run to over twenty-five papers.

For a short time Simpson returned to Paris to do further work with Wurtz, and later he spent a few years in London, where he was examiner to many public services, such as the Indian Civil Service, and to the Queen's University in Ireland. In 1872, at the age of 57, he went to Queen's College Cork as professor of chemistry. During his twenty years in Cork, Simpson devoted himself almost entirely to his teaching duties. He published only two further research papers.

Further reading:
Royal Society Year Book (London, 1903), 252–8.
Obituary Notice in the *Journal of the Chemical Society: Transactions* **81** (1902), 631–5.

William Davis
Department of Chemistry
Trinity College
Dublin

Born: Dublin, 1817

Died: Antrim, 3 January 1869, from scarlet fever

Family: Descendants of Huguenot refugees from Provence, who settled in Ireland about three hundred years ago. Son of Louis Victor du Noyer, music teacher.
    Married: Miss Du Bedat
    Children: Five. His eldest daughter predeceased him by one day.

Distinctions:
Honorary life member of the Royal Irish
    Academy 1863

Addresses:
1845	Seafort Avenue, Sandymount
1862–1863	Dunloe, Frankfort Avenue, Rathgar
1863–1867	Albert Ville, Sidney Avenue, Blackrock

George Victor du Noyer used his pencil, pen and pallet to record natural features and archaeological sites of the Irish countryside, much as the modern field worker uses a camera. He also prepared many important illustrations of fossils and carried out detailed geological surveys of large areas of Munster, Leinster and Ulster, as well publishing numerous scientific and archaeological papers.

Educated at Mr Jones's school in Dublin's Great Denmark Street, du Noyer studied art under George Petrie (1790–1866), the romantic landscape painter and pioneer of serious archaeological studies in Ireland. Petrie probably arranged for the young du Noyer to join the archaeologists, folklorists and natural historians employed by the Ordnance Survey during the 1830s in a regrettably short-lived venture to document the minutiae of the Irish countryside.

From 1835 until 1842 du Noyer was engaged in illustrating botanical, geological and zoological specimens for the Ordnance Survey's field parties, as well as in preparing many topographic pen and ink drawings. The four fine plates of grasses included in the Ordnance Survey's so-called 'Templemore Memoir' were selected for publication from the dozens of superb botanical drawings and watercolours by du Noyer that are now preserved in the National Botanical Gardens at Glasnevin. The hundreds of fossil illustrations, gathered into 38 plates in the *Report on the geology of Londonderry and parts of Tyrone and Fermanagh* by Joseph Ellison Portlock (1794–1864) are all by du Noyer and include scientifically very important records of the type specimens of individual species.

Made redundant by the Ordnance Survey, du Noyer taught art at St Columba's College, then at Stackallen, Co. Meath, before joining the staff of the Geological Survey of Ireland in 1846, the year after its foundation. Du Noyer had been trained by Portlock in making geological sections, and his first assignment in the Geological Survey was to record the geological structure exposed in the cuttings then being excavated for construction of the railways northwards and westwards from Dublin. As records of temporary exposures of the solid rock, du Noyer's railway sections have proved to be of lasting value, particularly in areas of thick glacial deposits where there are few natural surface showings of the solid rock.

Watercolour illustration by G.V. du Noyer of the rocks at Waterford Harbour showing flat-lying layers of Old Red Sandstone resting, unconformably, on steeply dipping, Lower Palaeozoic slates.

The railway sections completed, du Noyer joined the ranks of the geologists who were systematically mapping-out Ireland's geological structure. He surveyed all, or parts of, 42 of the 205 one-inch to the mile (1:63,360) geological map sheets that cover Ireland, and he contributed to seventeen of the accompanying 'explanatory memoirs', as well as constructing eight of the longitudinal geological sections published by the Geological Survey during his lifetime. Our present knowledge of the geological structure of large areas of counties Kerry, Cork, Tipperary (south riding), Waterford, Wexford, Carlow, Wicklow, Dublin, Meath, Louth, Cavan, Longford, Westmeath and Antrim owes much to du Noyer's work.

As well as official publications, du Noyer published numerous papers in learned journals. Of particular interest are his contributions to the contemporary debate on the origin of the so-called 'drift' deposits, because the papers include a fine account of many important glacial features discovered by du Noyer among the mountains of south Munster. A compulsive artist, du Noyer decorated his geological worksheets with sketches and watercolours. Many of the woodcuts that illustrate the Geological Survey's 'explanatory memoirs' or feature in contemporary textbooks are based on du Noyer's work, but the woodcuts scarely do justice to the beauty of the originals, many of which are preserved in the archives of the Geological Survey of Ireland.

No account of du Noyer's life would be complete without mention of his archaeological achievements, which included the original description of the *clochaun* or beehive hut settlement at Fahan on the Dingle Peninsula. Du Noyer presented eleven volumes of his superb drawings of abbeys, castles and prehistoric sites to the Royal Irish Academy and, shortly after his untimely death from scarlet fever, the Royal Society of Antiquaries in Ireland acquired twelve other volumes of his architectural and archaeological drawings.

Further reading:
Mary J.P. Scannell and Christina I. Houston, 'George V. du Noyer: a catalogue of plant paintings at the National Botanic Gardens, Glasnevin, with aspects of his scientific life'. *Journal of Life Sciences Royal Dublin Society* **2** (1980) 1–13.
Gordon Herries Davies, *Sheets of many colours: the mapping of Ireland's rocks 1750–1890* (Dublin, 1983). Royal Dublin Society Historical Studies in Irish Science and Technology No. 4.

Jean Archer
Geological Survey of Ireland
Dublin

Portrait of George Salmon. (Courtesy of the provost, fellows and scholars of Trinity College Dublin.)

Born: Cork, 25 September 1819

Died: Dublin, 22 January 1904

Family: Son of Michael Salmon, a Cork linen merchant, and his wife Helen (née Weekes). He had three sisters.
Married: Francis Anne Salvador
Children: Four sons and two daughters

Addresses:
1888-1904   Provost's House, Trinity College

Distinctions:
Fellow of Trinity College 1841
Member of the Royal Irish Academy 1843
Cunningham Medal of the Royal Irish
    Academy 1858
Fellow of the Royal Society, London 1863
Regius professor of divinity, University of
    Dublin 1866
Royal Medal of the Royal Society, London
    1868
Fellow of the Accademia dei Lincei, Rome
    1885
Provost of Trinity College 1888–1904
Copley Medal of the Royal Society,
    London 1889
Fellow of the British Academy 1902
Honorary degrees from Oxford,
    Cambridge, Edinburgh, Christiania
Honorary member of the Berlin,
    Göttingen and Copenhagen academies

George Salmon entered Trinity College Dublin at about the same time as James MacCullagh (see p. 22) was appointed to the chair of mathematics, so it is not surprising that his interests were led towards geometry where most of his significant mathematical contributions were to be made.

Unlike MacCullagh and Hamilton, Salmon had no interest in physics; he pursued mathematics for its own intrinsic interest. Salmon also differed from MacCullagh in his readiness to undertake formidable calculations: in this he resembled Hamilton (Vol. 1, p. 36), who seemed almost to enjoy heavy calculation for its own sake, whereas MacCullagh would prefer to devote long hours to searching for an elegant and simple way to reach a result. Hamilton tried but failed to interest him in quaternions.

Salmon's researches did extend into algebra, where he collaborated with the English mathematicians Cayley and Sylvester in pioneering work on matrices and the theory of groups. In fact this work in algebra, which was particularly concerned with the determination of invariants, was very relevant to geometry. The theory of invariants was subsequently used by Felix Klein in his so-called Erlangen Programme as a basis for a systematic description of geometry. Klein, who learned about invariants from a German translation of Salmon's book, was to meet his mentor in 1892 when he represented his university at the tercentenary celebrations in Trinity College at which Salmon presided as provost. To Klein's disappointment the provost no longer wished to talk about mathematics.

Although his original research contributions were substantial, Salmon's fame in mathematics stems primarily from his four textbooks: *Conic sections, Higher plane curves, Lessons introductory to the*

The Provost's House, Trinity College, Dublin. (Courtesy of the provost, fellows and scholars of Trinity College Dublin.)

modern higher algebra and *Geometry of three dimensions.* Written with elegance and clarity, these books enjoyed great success—they appeared in many editions and were translated into several languages. *Conic sections* remained in use as a textbook until comparatively recent times and it would not be altogether surprising to see a copy of one of the others on the desk of a research mathematician today.

As was normally required of a fellow of Trinity College at that time, George Salmon was ordained a priest in the Church of Ireland. In 1866, he was appointed regius professor of divinity and his main interest switched from mathematics to theology. He was a respected New Testament scholar, but the best known of his theological writings is *The infallibility of the Church,* a trenchant criticism of papal infallibility, which reflects its author's deep distrust of any authority over the individual conscience. In the reorganisation of the Church of Ireland which followed its disestablishment in 1870 Salmon played a central and constructive part.

In 1888 George Salmon was appointed provost of Trinity College and remained head of the college until his death in 1904. He was highly respected at home and abroad and he presided with much dignity over the college's tercentenary celebrations in 1892.

Further reading:
Entries in the *Dictionary of national biography* and the *Dictionary of scientific biography*
Obituary notices: *Proceedings of the British Academy* **1** (1903–4), xxii; *Proceedings of the London Mathematical Society*, 2nd ser., **1** (1903–4), xxii; *Proceedings of the Royal Society* **75** (1905), 347; *Nature* **69** (1903–4), 324.
T. D. Spearman, 'Mathematics and theoretical physics'. In T. Ó Raifeartaigh (ed.) *The Royal Irish Academy, a bicentennial history, 1785–1985* (Dublin, 1985).

David Spearman
School of Mathematics
Trinity College
Dublin

Portrait by Sarah Purser. (Courtesy of the provost, fellows and scholars of Trinity College Dublin.)

Addresses:

1847–1874	17 Heytesbury Terrace (afterwards 51 Wellington Road), Dublin
1875–1886	31 Baggot Street Upper, Dublin
1887–1897	12 Northbrook Road East, Dublin

Born: Carlow, 21 December, 1821

Died: Dublin, 31 October 1897

Family: The Haughtons were a Quaker family, although Samuel's parents were apparently not practising members of the Society of Friends and he had no difficulty in being ordained into the ministry of the Church of Ireland. Samuel Haughton and his wife Louisa had four sons.

Distinctions:
Fellow of Trinity College 1844
Member of the Royal Irish Academy 1845
Cunningham Medal of the Royal Irish Academy 1848
Professor of geology and mineralogy in the University of Dublin 1851-81
Fellow of the Royal Society, London 1858
President of the Royal Zoological Society of Ireland 1885–90
President of the Royal Irish Academy 1886–91
Honorary degrees from Oxford, Cambridge, Edinburgh, Bologna

Samuel Haughton has been appropriately described as a Victorian polymath. It was more common in Victorian times than it is today for a scholar or scientist to possess a broadly-based expertise but, even by the standards of his time, the range of Haughton's knowledge and the variety of fields to which he made original contributions were quite unusual. Mathematics was his first interest; MacCullagh's lectures (see p. 22), which he attended as an undergraduate at Trinity College Dublin, made a big impact. Later after MacCullagh's death, Haughton, with J.H. Jellett, published the collected papers of their former professor.

The problems which interested Haughton were usually of a practical type, preferably amenable to a quantitative description. One continuing field of interest was elasticity, what would today be called continuum mechanics. In 1848, four years after his election to fellowship in Trinity, he was awarded the Cunningham Medal of the Royal Irish Academy for his memoir *On the equilibrium and motion of solid and fluid bodies.* Haughton also shared MacCullagh's interest in the refraction of polarised light by crystalline media, which led him naturally to mineralogy. It was probably this, set against the background of an enthusiam for natural history that he had developed as a schoolboy in Carlow, that caused geology to take over from mathematics as his primary interest and, in 1851, he was appointed to the chair of geology. Haughton's geological work was wide-ranging. He did pioneering work on the chemical composition of rocks and also applied his mathematics to such topics as tidal motion and planetary equilibrium. He compiled comprehensive tide tables, which he applied to such varied topics as the incidence of shipwrecks, the sequence of events at the Battle of Clontarf, and the evidence which had been presented at a murder trial some years previously. This last study was one of several investigations

Sir Patrick Dun's Hospital when completed in 1816.

of a forensic nature which attracted his curisoity. Haughton was a pioneer in the geological study of climatic change and, in this context, he studied solar radiation and examined the effect of ocean currents on climate.

It was apparently his study of fossils which gave rise to an interest in anatomy and in 1859—the year in which Darwin published 'The origin of species'— Haughton, by now a fellow of the Royal Society, entered the Trinity medical school where he studied for three years, while still retaining his geology chair, before graduating M.B., M.D. in 1862. He carried out research into a variety of physiological problems, still favouring those which could be treated mathematically. He investigated muscular action and the mechanism of joints, publishing his results in 1873 as a book *The principle of animal mechanics.*

Following his election to fellowship Haughton took holy orders in the Church of Ireland. His unwavering opposition to Darwin's evolutionary theory was probably due both to religious and scientific conviction. Apart from his research and teaching, he played an active administrative role—in Trinity where he was registrar of the medical school, in the Academy where he served on Council for many years and was president from 1886 to 1891, on the board of Sir Patrick Dun's hospital, which he served for 35 years, and as honorary secretary of the Royal Zoological Society of Ireland. The restaurant building at the zoo, still known as the Haughton House, was built in his memory by public subscription. One other facet of this remarkable man, linked to his pedagogical interests, was the publication in joint authorship with his colleague J. A. Galbraith of a popular series of scientific manuals covering a wide range of topics.

Further reading:
W. J. E. Jessop, 'Samuel Haughton: a Victorian polymath'. *Hermathena* 116 (1973), 5.

David Spearman
School of Mathematics
Trinity College
Dublin

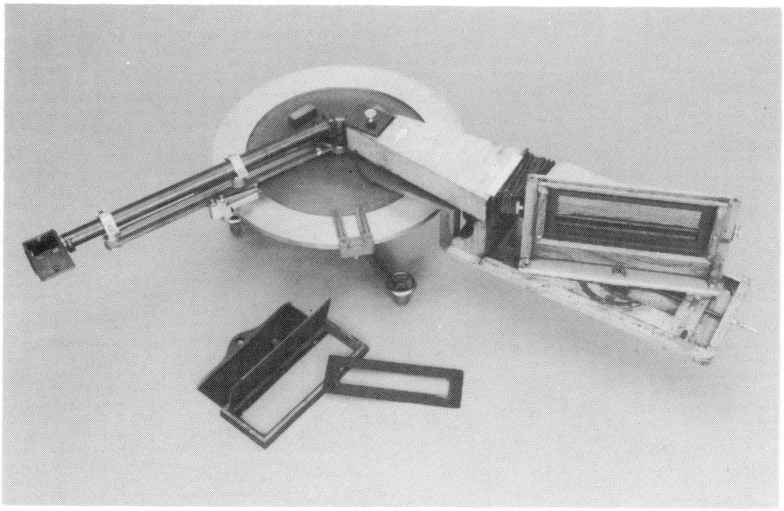

*Hartley's ultra-violet spectroscope: the prism table is signed ' Yeates & Son Dublin'.
(Courtesy, The Science Museum, London.)*

Addresses:
1883–1888 2 Tobernia Terrace, Monkstown,
              Co. Dublin
1889–1907 36 Waterloo Road, Dublin
1908–1912 10 Elgin Road, Ballsbridge, Dublin

Distinctions:
Professor of chemistry, Royal College
  of Science for Ireland 1879-1911
Fellow of the Royal Society 1884
Longstaff Medal of the Chemical Society 1906
Knighted 1911
Honorary D.Sc. Royal University of Ireland 1912

Born: Lichfield, England, 3 Febuary 1846

Died: Aberdeenshire, Scotland,
  11 September 1913

Family: Son of Thomas Hartley,
a portrait painter, and Caroline Lockwood
    Married: Mary Laffan of Blackrock,
Co. Dublin, a novelist and author 1882
    Children: His son W. J. Hartley became
a lecturer in agricultural bacteriology
in University College, Cardiff.

When Walter Hartley went to Edinburgh University he intended to study medicine, but he abandoned it for chemistry. His chemical education was spread over the years 1864–71 in Germany, Manchester and London. During the later part of this period he worked in the Royal Institution, where he performed the first research for which he was to become well-known. A French chemist, Bastian, said that he had created a living organism by heating for several hours a solution of inorganic chemicals (sodium phosphate and ammonium tartrate); in other words, he claimed to have achieved the spontaneous generation of life from lifeless matter. If true, this was an astonishing result. With brilliant experimental technique, taking particular care to prevent air from coming into contact with the solution, Hartley demonstrated that it was extremely unlikely that spontaneous generation had taken place, and far more probable that living organisms had entered the solution from the atmosphere.

In 1871 Hartley became senior demonstrator at King's College, London, and there he began his career as a spectroscopist. After moving to Dublin in 1879 to the Royal College of Science for Ireland, he took on heavy teaching duties, and most of his research was carried out during the college's holidays. The college premises were then situated in St Stephen's Green.

The science of spectroscopy had been created by the great German chemists Bunsen and Kirchhoff in 1859–60. It immediately became an important technique, but there was little understanding of the relationship between the lines in the spectra of different elements. Hartley was the first to establish that relationships do indeed exist between the wavelengths of spectral lines of the elements related to their positions in the classification called the Periodic Table. He published his results in 1883, though the discovery is usually associated with the names of Balmer (1884) and Rydberg (1885), who produced more mathematical formulations. This was perhaps Hartley's most important discovery: the detailed understanding of spectra was one of the paths which led to Quantum Theory.

Over the three decades in which Hartley worked in Dublin the main topic of his investigation was the relationship between the structure and the spectra of a wide variety of organic compounds. The importance of this subject had first been noticed by Sir George Gabriel Stokes, professor of mathematics at Cambridge and son of the rector of Skreen, Co. Sligo. Like so many of the scientists who worked in Ireland in the nineteenth century, Hartley was concerned with the practical application of scientific research. Typical of his work in organic spectroscopy was a study of dyes of particular relevance to the Irish textile industry. Derivatives of coal tar, such as benzene and naphthalene, could be converted into dyes with brilliant colours, even though benzene and naphthalene were themselves colourless. Hartley explained that the colours of the dyes were caused by the vibrations within the dye molecules: a molecule that could vibrate with a particular frequency could absorb light of that frequency (our present-day interpretation of these 'vibrations' is the movement of electrons between atomic energy levels. The electron was unknown when Hartley began his spectroscopic work.) He also pointed out that the apparently colourless benzene and naphthalene were in fact absorbing radiation with a higher frequency (or shorter wavelength) than dyes—in other words, they were absorbing the ultra violet. Hartley's other industrial research covered subjects such as brewing, distilling, the thermochemistry of the Bessemer process of steelmaking, and chemicals that might cure potato blight.

Although spectroscopy was the single method which Hartley deployed in almost all of his researches, he was interested in applying it to the whole natural world. Like that great Ulsterman, Lord Kelvin, he called himself a naturalist. The Scottish chemist and oceanographer J.Y. Buchanan sent Hartley fragments of shell from the ocean bed. Preserved in spirit, green colouring matter in the shells transferred itself to the spirit. Hartley demonstrated that the colouring was chlorophyll. He subsequently showed that a deep-sea animal, the sea cucumber, contains a variety of chlorophyll. These observations explained the intensely green colour of the very cold water which is found on certain ocean shores in the tropics: the cold water was identified as having come from the depths of the ocean. It had flowed to the coast to balance the warm surface water being blown off shore by trade winds.

Some of the spectroscopic apparatus used by Hartley is preserved at St Patrick's College, Maynooth. The Science Museum, London, has an ultraviolet spectroscope (see figure), made in part by Yeates of Dublin (see p. 72), which was presented by Hartley's widow in 1913.

Further reading:
J.Y. Buchanan ,'Obituary of W.N Hartley'. *Journal of the Chemical Society* **105** (1913), 1207–16.
W.N Hartley, 'On the colouring matters employed in the illuminations of the "Book of Kells"'. *Scientific Proceedings of the Royal Dublin Society* **4** (1885), 485–90.

John Burnett
National Museums of Scotland
Edinburgh

Courtesy National Library of Ireland.

Born: Raheny, Co. Dublin, 29 July 1847

Died: Carrablagh, Co. Donegal, 7 August 1908

Family: Third son of Sir Andrew Searle Hart, mathematician and vice-provost of Trinity College Dublin
Married: 1. Edith Donnelly, at Swords
(d. 1901), 1887
2. Mary Cheshire 1907
Children: two daughters by his first wife

The family's native county was Donegal — 'where my family has been settled since Elizabethan times'.

Distinctions:
Naturalist to the British polar expedition on
H.M.S. Discovery 1875–76
Fellow of the Linnaen Society, London 1875,
(rescinded 1878, reinstated 1887)
Assistant naturalist to the Palestine
exploration expedition 1883–84
Member of the Royal Irish Academy 1895
High sheriff of Donegal 1895
Fellow of the Royal Geographic Society 1898

Addresses:
1847   Glenvar, Raheny
Glenalla, Letterkenny, Co. Donegal
Carrablagh, Portsalon, Co. Donegal

Henry Chichester Hart was educated at Portora Royal School, Enniskillen, and Trinity College Dublin. He is known primarily as a botanist, but was distinguished in three spheres of activity — physical, scientific and literary. Described as a man of 'magnificent physique', his physical prowess and stamina were acknowledged by contemporaries.

Hart is prominent amongst those who laid the foundations of plant distribution studies in Ireland. Both Hart and R.M. Barrington were part of a small band of able botanists whom A.G. More marshalled for the purpose of exploration in connection with the 1898 edition of *Cybele Hibernica*. From an early age Hart took an interest in flora, influenced probably by his mother, who 'delighted in gardens and wild flowers'. At 'about seventeen years of age', he began the study of the plants of his native county. It was an ongoing project, culminating in the publication in 1898 of the *Flora of the County Donegal*. In the intervening years several papers were published, such as: 'Flora of north-west Donegal', 'Flora of the Croaghgorm range', 'Report on the flora of south-west Donegal', 'Botanic excursions in west Donegal'. The *Flora of the County Donegal*, an early county flora, set a pattern for those that followed. It is unusual, however, in that a sixth of the volume is taken up with extensive climatic data. Hart, who had a deep interest in climate, maintained meteorological equipment at his Donegal residence and was critical of the *Meteorological atlas of the British Isles* (1883) for the 'very slender knowledge' it provided for Ireland.

Hart's *Flora* is also noteworthy for an appendix setting out plant names and plant-lore accumulated during his travels throughout the county, in the course of which he exercised his ear for dialect. Praeger, in his monumental work *Irish topographical botany* (1901), reported that, for information on the north-west of the country, 'I rely almost entirely on the *Flora of Donegal*'.

It was but natural that Hart, with his scientific training and immense vitality, should seek to journey elsewhere in quest of plants. Aided by grants from the Royal Irish Academy, he surveyed islands, river valleys, stretches of coast-line and several mountain ranges. In 1875 he published 'A list of the plants found on the islands of Aran, Galway Bay'. He noted 372 species, of which 176 had not been previously reported. He also provided 'unambiguous data for the distribution of species on the three islands'. In 1880 Hart wrote 'On the botany of the Galtee Mountains...', in which he paid particular attention to the saxifrages, contrasting them with those specimens he had found on Aranmore, Co. Donegal. In 1882 'On the botany of the River Suir' was published and in 1883 'Notes on the flora of the Mayo and Galway mountains'. In 1890 he summed up his knowledge of upland plants in a paper 'On the range of flowering plants and ferns of the mountains of Ireland'. The report was the result of many years of field work. In it Hart set out the altitudinal ranges of alpine species in western mountains and in the Galtees, the Mournes and the Wicklow range. He reported new locations for alpines and a great deal of altitudinal data for plants of all types.

Hart also botanised outside Ireland — in the Arctic, in Sinai and in Palestine. Polar and Palestine plants collected by him are in the herbaria of Kew, the British Museum (Natural History) and the National Herbarium at Glasnevin. His specimens from Donegal and the mountain ranges of Ireland are in the National Herbarium. Some fossil material and a number of fish and invertebrates are in the National Museum of Ireland. A species first noted in west Donegal by Hart has been named in his honour *Saxifraga hartii* D. A. Webb (see figure). As a botanist Hart had great flair and ability. One notes in his papers the ease with which he assessed plant performance and distribution, and climatic and altitudinal data, while keeping in mind the folklore aspect of plants. He also made a notable contribution to zoological studies, publishing papers on mammals, insects, molluscs and birds.

This stamp, designed by Frances Poskitt, was issued by An Post in 1988 in the series Endangered Species of Vegetation. (Courtesy of An Post.)

In 1899 Hart read a paper to the Philological Society 'Notes on the Ulster dialect, chiefly Donegal'. Valuable manuscript notes are in the Royal Irish Academy. He was an authority on the works of Shakespeare and spent his later years editing the Arden series of plays — 'Othello', 'Merry wives of Windsor', 'Love's labours lost' and others.

Hart died at his home Carrablagh on 7 August 1908. His friend R. M. Barrington reported that he was buried at Glenalla in a spot chosen by himself amidst the beautiful scenery and surroundings of his ancestral home.

In the presentation of this article I received help from G.V. Hart, S.C., Michael Hewson, National Library of Ireland, Gina Douglas, The Linnaean Society of London, Alan Eager, Royal Dublin Society, Michael Mitchell, University College Galway, and Eric Pembrey, Dublin.

Further reading:

R.M. Barrington, 'Henry Chichester Hart'. *Irish Naturalist* **17** (1908), 249–54.

H.C. Hart, 'On the botany of the River Suir'. *Scientific Proceedings of the Royal Dublin Society* **4** (1883) 63–73.

H.C. Hart, Report on the flora of the Wexford and Waterford coasts'. *Scientific Proceedings of the Royal Dublin Society* **4** (1885), 117–46.

H.C. Hart, *The Flora of Howth* (Dublin, 1887).

M. Traynor, *The English dialect of Donegal — a glossary incorporating the collections of H.C. Hart* (Dublin, 1953).

Mary J.P. Scannell
National Botanic Gardens
Glasnevin
Dublin

Born: Youghal, 10 September 1847

Died: West Cove, Co. Kerry, 22 April 1919

Family: Only son of Charles Green, J.P., of Youghal
Married: Belinda Beatty, daughter of a fishery owner in Waterville, 22 June 1875
Children: Five daughters and one son, Charles Green (1876-1938), inspector of fisheries

Distinctions:
Fellow, Royal Geographical Society 1886
Honorary Member, Royal Dublin Society
  1888
Member, Royal Irish Academy 1895
Companion of the Bath 1907

Addresses:
1874–1877 Kenmare, Co. Kerry
1877–1889 Carrigaline, Co. Cork
1889–1914 5 Cowper Villas, Dublin
1914–1919 West Cove, Co. Kerry

William Spotswood Green was the founder of modern fisheries research in Ireland. Born in Youghal, he developed a love for the sea and fishing as a child. In 1859 he entered Rathmines School in Dublin, where the headmaster, C.W. Benson, was an enthusiastic naturalist. A school leaver in the 1860s, however, had no prospect of formal training or of a career in biology.

While an undergraduate in Trinity College Dublin, he made the first of a number of expeditions to study glaciers in the Alps and in Norway. Ordained in the Church of Ireland ministry, Green's first curacy was in Kenmare, after which he moved to Carrigaline. Both were coastal parishes and he established himself as an expert in marine science. In 1881 Green was advised, on grounds of ill health, to spend the winter away from Ireland. His reaction was characteristic of his phenomenal courage and energy. He set off for New Zealand to climb Mount Cook, an unconquered peak with unexplored glaciers.

Green's formal connection with marine research began in 1885 when the Royal Irish Academy initiated studies on the continental shelf off the south-west coast. His reputation as an expert on sea fishing and boat handling led to an invitation to organise the first expeditions. The steamer *Lord Bandon* was chartered for short periods in the summers of 1886 and 1887 and carried out dredging and trawling down to 2000 metres. The deepest dredging was a failure, but faunal specimens were collected down to 594m. They included a sea anemone new to science, which was named in his honour *Paraphellia greenii*.

The Royal Dublin Society in 1887 commissioned Green to make a report on the state of the fishing industry. This was to be the groundwork for a scheme to develop the fisheries and it led directly to a series of exploratory cruises, including one more funded by the Royal Irish Academy, which took place in May 1888. The *Lord Bandon* was chartered again, but under the new name of *Flying Falcon*. This time successful hauls were made down to 2322m.

Later in the same year Green set off for Canada to study the Selkirk Glaciers in the Rocky Mountains. After the expedition he made a study tour of North American fishing ports on behalf of the Royal Dublin Society. His observations led to important innovations in the Irish fishing industry.

His next expedition was one mounted by the British Museum to collect deep sea fish off the south-west coast in 1889. This was Green's final voyage as a country clergyman. The following year, he was appointed one of the three inspectors of Irish fisheries.

Marine exploration continued, this time jointly sponsored by the Royal Dublin Society and the government. In 1891 Green not only organised the expedition, but skippered the steam yacht *Harlequin* from Southampton to Cobh, up the west coast and round to Dublin. A team of scientists took part in the voyage and their reports form the basis of scientific information on marine life in Irish coastal waters.

As inspector of fisheries, Green continued to organise research cruises. The most spectacular were two voyages to Rockall in June 1896 that yielded valuable scientific data in spite of foul weather. During these years, Green supervised the work of a number of marine scientists and eventually, in 1900, a permanent government research body was set up in conjunction with the Department of Agriculture and Technical Instruction for Ireland. In that year he was appointed chief inspector of fisheries. Under his supervision, the Department purchased its own vessel for research and patrol, the first of two ships named *Helga*. The post of chief inspector was partly administrative, partly scientific, and established the necessity for a biological basis in national fisheries management.

Green retired in 1914 to a seaside village in Kerry where he died in 1919. He was buried in Sneem, in sight of the ocean. He published scientific papers on mountain glaciers and marine archaeology besides his fisheries work. R.L. Praeger, who went on several of the cruises, wrote with deep affection of Green as a man whose dedication to work was equalled only by his sense of humour and the desire to help others.

Further reading:
A.E.J. Went, 'William Spotswood Green'. *Scientific Proceedings of the Royal Dublin Society* **B2** (1967), 17–35.
R.L. Praeger, *The way that I went* (Dublin, 1939).

Christopher Moriarty
Department of the Marine
Abbotstown
Dublin

Dr Henry's portrait in the Forestry Herbarium at the National Botanic Gardens, Glasnevin, Dublin; on the table is a copy in original wrappers of Elwes' and Henry's seven-volume Trees of Great Britain and Ireland.

Born: Dundee, Scotland, 2 July 1857

Died: Dublin, 23 March 1930

Family: Son of Bernard and Mary (née McNamee) Henry of Tyanee, Portglenone, County Derry
  Married: 1, Caroline Orridge, 20 June 1891: she died in Denver, Colorado, in September 1894;
  2, Alice Brunton, 17 March 1908: she died in Dublin in 1956.
  There were no children.

Distinctions:
Fellow of the Linnaean Society 1888
Veitch Memorial Medal 1902
Victoria Medal of Honour 1906
Silver Medal, Société Nationale
  d'Acclimatation de France 1923

Augustine Henry was born in Scotland, but his parents soon returned to the family home, near Portglenone, on the west side of the River Bann in County Derry. There Augustine spent his childhood. He was a bright lad, leaving school (Cookstown Academy) to enter Queen's College (now University College) Galway, where he studied natural history and medicine, gaining his B.A. in 1877. Shortly after completing his master's degree at Queen's College Belfast, Henry was persuaded by Sir Robert Hart, the Ulster-born diplomat and civil servant, to join the Chinese Imperial Maritime Customs Service as a doctor. In the summer of 1881, aged 24, Augustine Henry set sail for China: he was to live in remote parts of that vast country for most of the succeeding twenty years.

At first he was stationed at Yichang, a port situated 1000 miles inland from Shanghai on the Chang Jiang (River Yangtze), where he treated the sick and also undertook duties as a customs officer. Henry soon became fascinated by the native medicines and it was this interest that led him to undertake the collection of plant specimens for study and identification at the Royal Botanic Gardens, Kew.

Because few Europeans had collected the plants of central China, Augustine Henry's specimens proved to be new and exciting — during his years in Yichang he discovered over five hundred new species, about twenty-five new plant genera, and one entirely new plant family. Many of his discoveries proved to be attractive, hardy plants suitable for European gardens, and Henry sent seeds and bulbs to Kew for cultivation there. Among his plant introductions were the orange-blossomed lily, *Lilium henryi*, which inhabits the limestone gorges near Yichang. Later he collected in southern China too, but his discoveries there are not suitable for cultivation even in the mildest parts of Ireland, and they are less familiar to us. Henry did collect in the field but he also employed native Chinese collectors to gather, dry and press specimens. The principal set of his Chinese plants is now in Kew.

He tried to encourage others to collect plants, especially those of economic and medicinal value, and published a small handbook to assist collectors. In it he predicted that the Chinese gooseberries (species of *Actinidia*) had considerable potential as an edible fruit—the misnamed 'kiwi fruit' is one of these.

After completing his time in China, Dr Henry took up the study of trees (dendrology), beginning by entering the famous French School of Forestry at Nancy. In 1903, Henry Elwes, a wealthy English landowner and gardener, invited Henry to collaborate In writing a book on the native and exotic trees cultivated in British and Irish gardens. The magnificent seven-volume book, illustrated with photographs, was completed in 1913, the same year that he was appointed professor of forestry at the Royal College of Science (later University College) Dublin. He was to remain there until he retired in 1927.

Augustine Henry promoted the idea of planting marginal land with fast growing coniferous trees, principally those from western North America. He is thus the 'father of Irish forestry'. He also did pioneering work on the production of vigorous artificial hybrid trees, especially poplars. His forestry herbarium, containing specimens cultivated in Ireland and Britain, is now at the National Botanic Gardens, Glasnevin.

Many botanical collectors are commemorated by having their names incorporated into the scientific names of plants that they found. Augustine Henry is no exception. There are too many to list but these plants mark his contribution to the discovery of the Chinese flora: *Rhododendron augustinii* (a blue-flowered rhododendron), *Parthenocissus henryana* (a red-leaved vine), *Acer henryi* (a maple), and *Hypericum augustinii* (a St John's wort). His first wife Caroline is also commemorated in the lovely primrose, *Carolinella henryi*.

Dr Henry was also an unwitting agent of a famous botanical hoax — a made-up plant concocted by one of his Chinese collectors fooled botanists at Kew and was named *Actinotinus sinensis*.

Principal publications:
Notes on economic botany of China (Shanghai,1893; facsimile reprint, Kilkenny, 1986).
With Henry J. Elwes: *The trees of Great Britain and Ireland* (Edinburgh, 1907–13).
Forests, woods and trees in relation to hygiene (London, 1919).

Sources and further reading:
S. Pim, *The wood and the trees, a biography of Augustine Henry* (2nd revised edition, Kilkenny, 1985).
E.C. Nelson, 'Some botanical hoaxes and Chinese puzzles'. *Kew Magazine* **3** (1987), 178–85.

E. Charles Nelson
National Botanic Gardens
Glasnevin
Dublin

Born: Farnworth, Lancashire, 29 December 1857

Died: Clifton, Lancashire, 7 April 1937

Family: Third son of a Liverpool merchant. His brother Alfred was a mathematician and was elected a fellow of the Royal Society in 1934
Married: Grace Martha Kimmins (1896), sister of educationalist Charles Kimmins
Children: One of his twin sons was killed at Ypres in 1915; the other became headmaster of Lincoln School and later of Rossall.

Distinctions:
Professor of chemistry, University College
    Bristol 1887–1903
Fellow of the Institute of Chemistry 1888
Fellow of the Royal Society 1893
Founder fellow of the Institute of Physics 1893
Professor of chemistry, Trinity College
    Dublin 1903–28
President of the chemical section, British
    Association 1904
President of the Royal Irish Academy 1921–26

Addresses:
1906–1916 12 Raglan Road, Ballsbridge, Dublin
1917–1930 13 Clyde Road, Ballsbridge, Dublin

Distillation as a method of concentration or separation of liquids has a very ancient history. Alchemical writings show that it has engaged chemists' attention from earliest times and its often central role in chemical processes continues in the present day. We owe much of our present knowledge to Sydney Young, a chemist who spent over half his professional life in Dublin.

Sydney Young received his chemical training at Owens College, Manchester, under the guidance of Sir Henry Roscoe. Later he spent some time working with the German organic chemist Rudolf Fittig in Strasbourg where, as well as learning organic chemistry, he acquired the glass-blowing skills which later he found so useful in his work on distillation. His first work as a student at Manchester was to investigate the discovery that ice cannot be liquefied by great heat so long as it is kept at low pressure under vacuum. (This finds modern utility in the process we now call 'freeze drying'. His Dublin colleague John Joly was to show that the inverse was true, that is, that ice melts at a lower temperature when under pressure.) So started an interest in the relationships between the states of matter, solids, liquids and gases which was to remain a continuing theme of his scientific researches.

Young was a pioneer in the separation and specification of pure organic chemicals. Changes of state or phase or of composition are mirrored by the heat changes that accompany them, the study of which is usually called thermochemistry. The rules of this behaviour are enshrined in the laws of thermodynamics, a subject that was being developed when Young was carrying out his early work. Thermodynamics makes

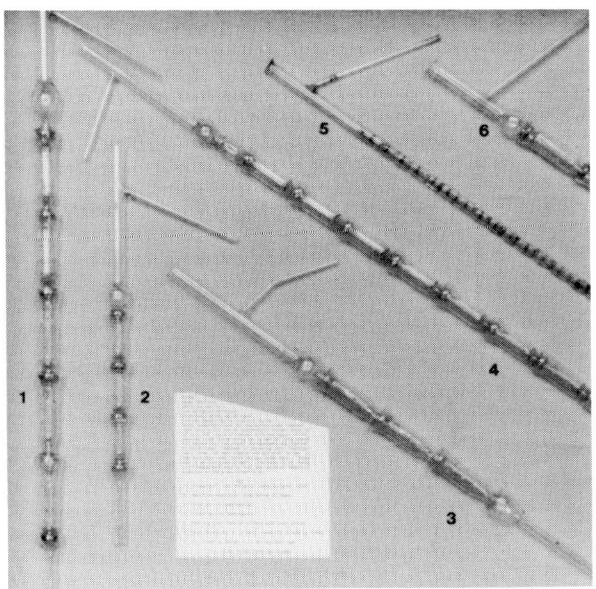

Glass fractionating columns made by Sydney Young (from a display in the Chemistry Department, Trinity College Dublin).

Petroleum Fractional Distillation Columns at Whitegate Refinery. The town of Cobh is in the background. (Courtesy of Irish Refining plc.)

evident the conditions under which separation of substances of similar boiling point can be carried out in a process called fractional distillation. By careful determinations of critical and other physical constants, and the relationship between temperature and vapour pressure, Young clarified crucial thermodynamical relationships for solids and liquids. From 1885, the potential for extracting paraffin hydrocarbons from petroleum led Young to devise improved techniques of fractional distillation. He developed a new method of dehydrating ethyl alcohol using benzene in an 'azeotropic mixture'; in 1902 he published this method in a paper entitled 'The preparation of absolute alcohol from strong spirit'. This was immediately before he came to Dublin in 1903.

At the beginning of the twentieth century Young was regarded as the foremost authority in the world on distillation. He continued his research in Dublin but, because of his increased teaching and administrative duties, he carried out scientific work at a lesser pace than before. He held a number of positions on government and scientific committees. He was a founder member of the Irish section of the Royal Institute of Chemistry. His text-book *Stoichiometry* and a new and enlarged edition of his book on distillation were published while he was professor in Trinity College. He also contributed articles on distillation, thermochemistry and thermometers to Thorpe's *Dictionary of applied chemistry*.

Further reading:
S. Young, *Practical distillation* (London, 1903).
S. Young, *Stoichiometry* (London, 1908; 2nd edition 1918).
S. Young, *Distillation principles and processes* (London, 1922).

William J. Davis
Department of Chemistry
Trinity College
Dublin

Portrait of Anderson in presidential robes by Charles Lamb, R.H.A.

Born: Coleraine, Co. Derry, 12 May 1858

Died: Dublin, 5 September 1936

Family: Son of Daniel Anderson of Camus, Coleraine
   Married: Emily Binns of Galway
   Children: Three daughters, Elisabeth, Emily and Helen; one son, Alexander

Addresses:
1891–1895 Tariffa Lodge, Galway
1895–1899 Sea Road, Galway
1899–1934 President's Residence, University College Galway
1934–1936 7 Pembroke Park, Dublin

Distinctions:
Queen's College Galway B.A.1880, M.A. 1881
Sixth wrangler, University of Cambridge 1884
Professor of natural philosophy Queen's College Galway 1885
Fellow of the Royal University of Ireland 1886
Fellow of Sidney Sussex College Cambridge 1891
President of Queen's College Galway 1899
Honorary LL.D. University of Glasgow 1901
Honorary D.Sc. National University of Ireland 1909
Member of the Royal Irish Academy 1909

Like many a northerner of his time Alexander Anderson sought his university education in Galway on the banks of the Corrib. In 1878 he began a distinguished career and an association of more than fifty years with that college. Having won scholarships in every year, he graduated in physics and mathematics with a B.A. and gold medal in 1880 and an M.A. in 1881. He then spent several years in Sidney Sussex College in the University of Cambridge studying physics and mathematics under J.J. Thomson before returning to Galway in 1885 to succeed Joseph Larmor as professor of natural philosophy. He held what became the chairs of experimental physics and mathematical physics until his retirement in 1934.

Shortly after his appointment as professor, practical physics classes were introduced into the curricula of science, medical and engineering students. This was in line with the trends of those times. However, it was not done without considerable trouble — for instance, the Board of Public Works, then in control of the college buildings, had to be approached five times before they would consent to alter the rooms to suit the new requirements. The college had acquired a considerable amount of scientific apparatus before then. An illustrated stock list of the apparatus held in 1861 showed that it possessed a wide variety of items for demonstrating various aspects of physics such as optics, electrostatics and mechanics. Other instruments were acquired over the years, and another list of apparatus, printed in 1902, showed that such recent developments as X-rays, cathode rays and Marconi's radio apparatus were already in use in the physics department in Galway.

Anderson's interest in the practical applications of physics is illustrated by the fact that the physics department of Queen's College Galway was providing a medical radiography service in Galway from 1898. He was also involved in industrially sponsored research, as the Kodak Company provided a fellowship for the study of X-ray photography in Queen's College Galway in 1900. This activity attracted

Physics Laboratory in Queen's College Galway about 1902.

world-wide attention in 1904 when the college authorities were taken to court in what was probably the first instance of litigation on the injurious effects of ionising radiations. A child had been brought for X-ray of a knee in which a piece of a needle was thought to be embedded. No part was found but the tissue was damaged and scarred by the X-ray exposure. Many well known physicists of the time, like Lord Kelvin and Sir Oliver Lodge, came to Dublin as expert witnesses at the trial. The verdict was in favour of the college and the child recovered without any permanent disability.

Professor Anderson's research covered a wide range of topics and was reported in some thirty articles, mainly in the *Philosophical Magazine.* This range may be illustrated by titles of a few of his papers: 'The focometry of diverging lens combinations' (*Phil. Mag.,* 5th ser., 31 (1891)); 'On the comparison of two self-inductions' (*Phil. Mag.,* 6th ser., 22 (1911)); 'On contact difference of potential and the action of ultraviolet light' (with H.N. Morrison, *Phil. Mag.,* 6th ser., 24 (1912)); 'On the method of measuring surface tension and the angles of contact' (with J.E. Bowen, *Phil. Mag.,* 6th ser., 31 (1916)), 'On scalar and vector potentials due to moving electric charges' (*Phil. Mag.,* 6th ser., 43 (1922)).

While the Anderson bridge method of measuring the self inductance of a circuit is probably his best known contribution to practical physics, his constant volume method of measuring the viscosity of a gas and the pendant drop method of measuring the surface tension of a liquid were also included in textbooks of practical physics, such as Worsnop and Flint, *Advanced practical physics* (London, 1957).

Besides his work of teaching and research in experimental and mathematical physics, Professor Anderson proved himself as an able university administrator. He was president of the Galway college from 1899 to 1934. During his term of office there were many new developments both in Ireland and in the position of the college — the incorporation in 1908 of the old Queen's College in the new National University of Ireland, the establishment of the Irish Free State and the beginning of the gaelicisation of the college being some of the more revolutionary changes. Throughout all these changes his tact and broadminded attitude avoided many a difficulty and solved many a problem. When, about 1925, the very existence of the Galway college was at stake, his great reputation and skilful direction were, to a large extent, responsible for averting the danger.

On his retirement he went to live in Dublin, where he died on 5 September 1936. He was buried in Dean's Grange Cemetery.

Further reading:
Obituary in *Nature* **138** (1936), 609.

Thomas O'Connor
Department of Physics
University College
Galway

Born: Ballyhagan, Kilmore, Co. Armagh, 23 July 1860

Died: Dublin, 7 March 1900

Family: Preston's wife Katherine, who survived him by more than half a century, became principal of Alexandra College. His son George was the co-discoverer of Guinier-Preston zones in metallurgy. He had two other children, one of whom died in infancy.

Distinctions:
Member of the Royal Irish Academy 1897
Fellow of the Royal Society 1898
Honorary D.Sc. 1898
Boyle Medal of the Royal Dublin Society
    1899, presented 8 February 1900

Address:
Bardowie, Orwell Park, Rathgar, Dublin

Thomas Preston enrolled in Trinity College Dublin in 1881, the year in which G. F. Fitzgerald (Vol. 1 p. 62) became professor, and he was one of the many who benefited from the great man's leadership in the period leading up to the end of the century, shortly after which both men suffered untimely deaths. This was the more so in Preston's case, for he was at the very height of his powers. But one need not speculate on mere potential: Preston recorded many outstanding achievements in the last decade of his life. These include the discovery of the Anomalous Zeeman Effect and the writing of two authoritative textbooks. *The theory of light* (1890) and *The theory of heat* (1894) are exceptionally thorough and professional treatments of their subjects. A wealth of informative detail is carried along by a magnificent prose style and beautiful illustrations. Preston produced an expanded second edition of *Light* in 1895 and the world wide popularity of both books was such that they survived as recommended texts throughout half of the present century, with further editions revised by other hands.

This work was begun while Preston was still associated with Trinity, and continued after he became professor of natural philosophy at University College Dublin in 1891. The post carried with it fellowship of the Royal University and, in due course, access to the new spectroscopic facilities which were installed by W. E. Adeney. Using these, together with a magnet borrowed from W. F. Barrett (Royal College of Science), Preston embarked on an investigation of the Zeeman Effect in 1897. In this he was undoubtedly encouraged by Fitzgerald's theoretical interest, but his investigations were as single-handed as they were single-minded. Despite competing demands on his time (he was, among other things, inspector of science and arts), he succeeded first in surpassing Zeeman's original observations and then in discovering that the true effect was more complicated: hence the term Anomalous Zeeman Effect. Preston's publication established priority over many other distinguished physicists who had been drawn to this fashionable

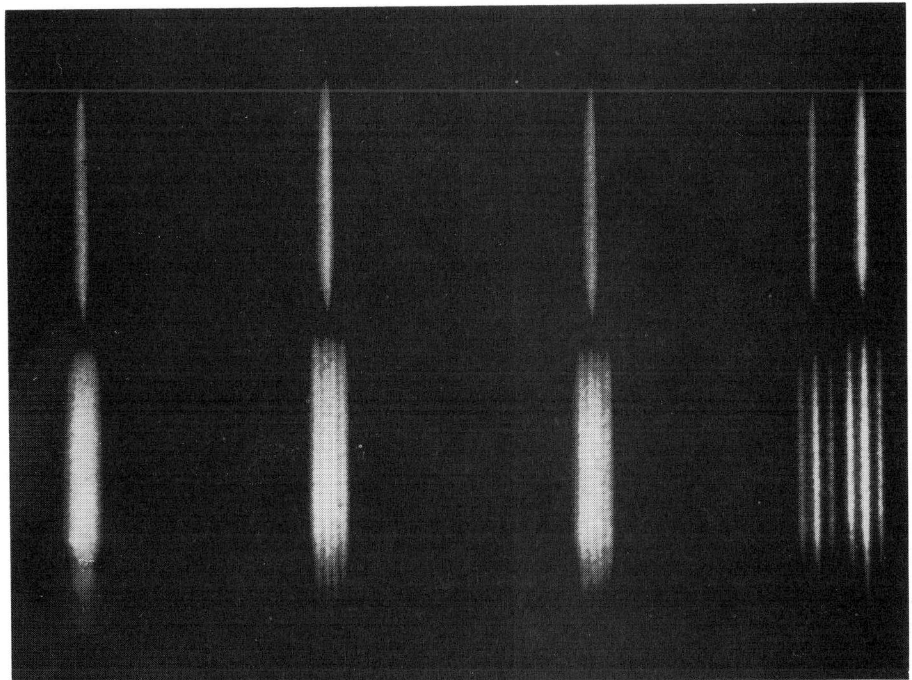

This photograph is taken from Preston's paper to the Royal Dublin Society in December 1897. It shows spectrum lines of cadmium and zinc photographed in zero magnetic field (top) and a strong magnetic field (bottom). The two lines on the right hand side show a triplet splitting of the type reported by Zeeman and anticipated by Lorentz's electron theory. The next two lines show a distinct quartet structure. This was the first reported example of the Anomalous Zeeman Effect. Preston's work demonstrated the inaccuracy of classical theory, and it was not until the 1920s that a proper explanation of Zeeman splitting was produced with the introduction of the concept of electron spin and the development of quantum mechanics.

topic. He went on to establish empirical rules for spectral lines, which are still associated with his name, and was still working in that area when he died of a perforated ulcer, presumably brought on by overwork.

Although much praised and honoured at the time, Preston's achievements have been insufficiently recognised since then. In particular it is noteworthy that Zeeman and Lorentz received the Nobel Prize for their pioneering work in the area in which Preston seized the lead so dramatically in 1897.

Further reading:
D. Weaire and S. O'Connor; 'Unfulfilled renown: Thomas Preston (1860–1900) and the Anomalous Zeeman Effect'. *Annals of Science* **44** (1987), 617–44.

D. Weaire
Department of Physics
Trinity College
Dublin

S. O'Connor
Department of Physics
University College
Dublin

Born: Ballymena, Co. Antrim,
31 January 1864

Died: Dublin, 27 April 1933

Family: Daughter of William Knowles of
Ballymena, an insurance agent, who had an
interest in natural history. She had a sister,
Catherine.

Addresses: Not known

'One of the finest pieces of work ever carried out in any section of the Irish flora' was how *The lichens of Ireland*, written by Matilda Knowles and published in 1929 by the Royal Irish Academy, was described. Matilda had been encouraged to study natural history by her father, an insurance agent by profession and in his sparetime an enthusiastic naturalist with a special interest in prehistoric man. He brought Matilda and her sister to meetings of the Belfast Naturalists' Field Club, which ran regular lectures, excursions and field trips. Through the field trips she first met Robert Lloyd Praeger, who became the best known Irish naturalist of his generation and her life-long friend. Between about 1896 and 1900, to further their education, both girls attended natural science classes at the Royal College of Science and Art in Dublin, since women were then unable to take degrees at Trinity College. In 1902 Matilda successfully started her chosen career as a botanist by obtaining a post in the National Museum of Science and Art (called the National Museum of Ireland from 1921). Five years later she became assistant to the head of the botany department, Professor Thomas Johnson. She eventually took complete charge of the botanical collections when he retired in 1923. Her meticulous and dedicated curatorship of the national herbarium, assisted by Margaret Buchanan, was reflected in the excellent condition of the collections achieved during her period of management when renowned botanists from many parts of the world corresponded with her, researched the collections and contributed numerous specimens. She published many scientific papers on a wide range of botanical subjects during her employment in the Museum. *The handlist of Irish flowering plants and ferns*, first published in 1910, went to three editions and is generally attributed to Professor Johnson, but was almost entirely written by her.

Lichens, in which Matilda Knowles developed a special interest, are a difficult group of plants. In 1908 she was invited to join the botanist Lorrain Smith, of the British Museum, to survey the lichens of Clare Island, Co. Mayo, which was to be part of a complete investigation of its natural history (Vol. 1, p. 94) in which notable naturalists from Ireland, Britain and Europe were invited to take part. In acknowledging her contribution to the survey she was described by Lorrain Smith as having 'exceptional ability as an observer and collector'. Subsequently Matilda Knowles devoted much of her time to the collection and study of lichens, working towards what was to be her *magnum opus*, *The lichens of Ireland*, which added over a hundred species to the Irish list and recorded the distribution of the 800 species found in Ireland.

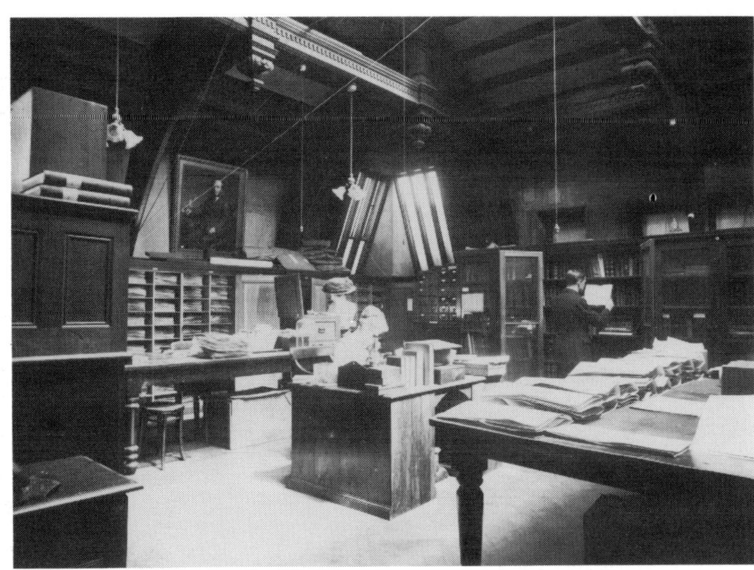

Matilda Knowles at her desk in the Museum about 1910. Courtesy of the National Museum of Ireland.

The Science and Art Museum, Kildare Street, Dublin, 1910.

It is apparent that her wide botanical knowledge, interest and enthusiasm enabled her to make an important contribution to Irish botany, although she did not receive any academic recognition in her lifetime other than from friends and acquaintances who held her in high esteem.

I am especially indebted to Miss Maura Scannell, National Botanic Gardens Glasnevin, for information on the botanical work of Miss Knowles, to Dr Jim O'Connor, National Museum of Ireland, and Mr Tim Collins, University College Galway, for additional information.

Further reading:
Praeger, R.L., 'Matilda Cullen Knowles'. *Irish Naturalists' Journal* **4** (1933), 191–3.
Ross, H.C.G., *Matilda Knowles — botanist*. Poster and teachers' notes published by the Ulster Museum and the Equal Opportunities Commission for Northern Ireland as part of a series on Irish women scientists, 1985.

Helena Ross
Ulster Museum
Belfast

Born: Dublin, 19 May 1869

Died: Dublin, 20 December 1953

Family: Son of George Dixon, soap-manufacturer, and Rebecca Yeates, daughter of Dublin's leading instrument-maker
   Married: Dorothea Mary Franks, daughter of Sir John Franks, C.B. 1907
   Children: three sons, two of them scientific fellows of King's College, Cambridge

Addresses:
1869–1880   30 Holles Street, Dublin
1881–1888   4 Earlsfort Terrace, Dublin
1889–1894   17 Earlsfort Terrace, Dublin
1894–1905   23 Northbrook Road, Dublin
1906–1907   Milverton, Temple Road, Dublin
1908–1934   Clevedon, Temple Road, Dublin
1934–1953   Somerset, Temple Road, Dublin

Distinctions:
Professor of botany, Trinity College Dublin
   1904–49
Fellow of the Royal Society 1908
Boyle Medal of the Royal Dublin Society 1917
Croonian lecturer, Royal Society 1937
President, Royal Dublin Society 1944–47
Member of the Royal Irish Academy 1947
Honorary fellow, Trinity College Dublin 1950
Honorary president, International Botanical
   Congress, Stockholm 1950

Dixon was the youngest son of a large and talented family; two of his brothers also became professors. He entered Trinity College in 1887 and won a foundation scholarship in classics in 1890, but then changed to natural science, in which he graduated with a gold medal in 1891. The change was made under the influence of John Joly (Vol. 1, p. 64), with whom he went on a walking-tour in Switzerland in 1888; this opened his eyes to the beauty and wonder of the natural world and laid the foundations of a life-long friendship.

From 1892 to 1904 Dixon was assistant to E.P Wright, whom he succeeded in the chair of botany in 1904, but he first spent a year at Bonn, where Strasburger was working out the details of the recently discovered process of meiosis. Dixon's studies here were mainly cytological, but his imagination was also stimulated by Strasburger's demonstration that the sap in trees could rise even when the cells in the stem had been killed. Back in Dublin he considered the implications of this. If the living cells in the wood did not pump up the sap (as had generally been supposed), then it must rise under purely physical forces, root-pressure, capillarity and atmospheric pressure all being plainly inadequate. With Joly's help he elaborated a theory that the motive force was primarily the evaporation from the leaves; this produces a tension in the water filling the vessels, and thereby, thanks to the cohesive strength of a bubble-free column of water (far greater than had been hitherto imagined), draws up the water from the roots. Many experiments had to be devised to refute sceptics; they were done with simple but ingenious apparatus and explained with patience and clarity. By about 1914 most of his critics had been silenced, and a few years later the cohesion theory found its way into the textbooks as undisputed orthodoxy.

The School of Botany, Trinity College Dublin.

Dixon had a very alert and imaginative mind, and he made many pioneering observations and suggestions which, because he did not follow them up, are often credited to others. These include the explanation of bivalent chromosomes at meiosis, the use of thermocouples for determining osmotic pressure in small volumes of fluid, and the mutagenic effect of cosmic rays. His theory of the transport of organic substances in plants turned out to be mistaken, but it stimulated research in a field in which much still remains to be explained.

In 1904 botany was still taught in cramped and unsatisfactory rooms originally designed for residence. Thanks to the persistent advocacy of Joly and the generosity of Lord Iveagh, money became available in 1906 for the erection of a proper laboratory. In the design of this building Dixon almost played as great a part as the architect, and with great success. It gave scope for the development of the subject by twentieth-century standards, and even today its only serious defect is that it is too small for the greatly increased numbers of staff and students. In 1910 an annex was added to house the herbarium adequately for the first time in its long history, and it is to Dixon's credit that, although he had no real interest in taxonomy, he devoted the best part of two years to filing and indexing the specimens.

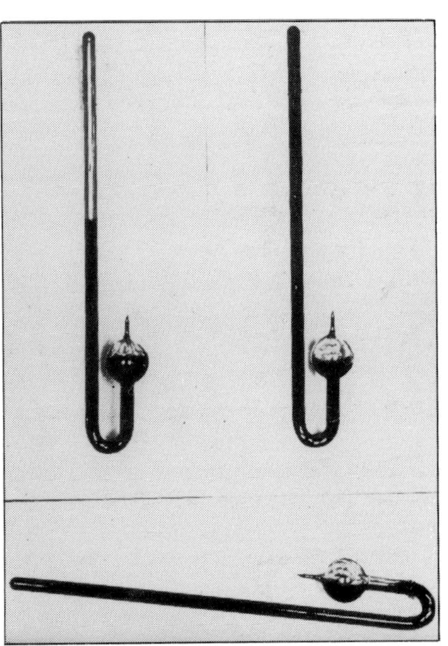

Apparatus devised by Dixon to demonstrate the cohesive strength of water. The J-tube contains only coloured water and water-vapour, and if inclined so as to fill the long arm (bottom) and then raised carefully the long arm (right) remains full.

Further reading:
H. H. Dixon, *Transpiration and the ascent of sap in plants* (London, 1914).
W.R.G. Atkins, 'Henry Horatio Dixon'. *Obituary notices of fellows of the Royal Society* **9** (London, 1954), 79–97

David Webb
School of Botany
Trinity College
Dublin

Born: Wexford, 2 October 1875

Died: Dublin, 11 July 1950

Family: Son of Myles and Teresa (née Harris) Conway
   Married: Agnes Christina Bingham, daughter of William Bingham of Ballymena, Co. Antrim, 1903
   Children: Teresa Mary, Morgan, Verna and Orlaith

Distinctions:
Professor of mathematical physics, University College Dublin 1901
Honorary D. Sc. Royal University of Ireland 1908
Fellow of the Royal Society, London 1915
President of the Royal Irish Academy 1937
Honorary Sc.D. University of Dublin 1938
Honorary LL.D. University of St Andrews 1938
Member of the Pontifical Academy of Sciences 1939
Honorary fellow of the Corpus Christi College Oxford 1940
President of University College Dublin 1940
President of the Royal Dublin Society 1942

Addresses:
1903–1912   100 Leinster Road, Rathmines, Dublin
1912–1920   Elsinore, Coliemore Road, Dalkey, Co. Dublin
1920–1929   Abbeyview, Coliemore Road, Dalkey, Co. Dublin
1929–1950   Colamore Lodge, Coliemore Road, Dalkey, Co. Dublin

Arthur William Conway entered University College, 86 St Stephen's Green, Dublin, in 1892 as a resident student and in 1897 he received the M.A. degree with first class honours in mathematical science. He then transferred to Corpus Christi College, Oxford, where in 1898 he was awarded the Junior and in 1902 the Senior University Scholarship in mathematics. In 1900 he became a junior fellow of the Royal University of Ireland. Appointed to the chair of mathematical physics at University College in 1901, he was also external lecturer at St Patrick's College, Maynooth, from 1903 to 1910, during the transition period when Maynooth was preparing for association with the National University of Ireland. He was succeeded in this lectureship by his student Eamon de Valera.

On the establishment in 1908 of the National University, with University College as one of its constituent colleges, Conway was appointed registrar of the college. He occupied this position until he became president of the college 32 years later. In spite of his teaching and administrative duties, Conway pursued scientific research until the end of his life, when he left behind papers that were published posthumously. His research was greatly influenced by the investigations of the nineteenth-century Irish mathematician Sir William Rowan Hamilton (Vol. 1, p. 36). This is seen by his interest in quaternions, on which he continued to work from 1900 until his last years, by when he had come to be acknowledged as the world's greatest authority on the subject. It is also shown by his undertaking and accomplishing the arduous task of joint editor of the first two volumes of Hamilton's mathematical papers, the first on geometrical optics being published in 1931 in collaboration with J. L. Synge, and the second on dynamics in 1941 in collaboration with A. J. Mc Connell.

In the course of a discussion on elementary particles at a meeting of the Royal Irish Academy about 1945, Conway recalled that during his scientific life the electron had been discovered by J.J. Thomson. This discovery was a crucial point in the history of theoretical physics. Conway was to follow closely developments in atomic models and to apply his mathematical dexterity in elucidating and developing the new theories of atomic structure that were emerging. In the middle of the first decade of the present century, it was generally supposed that the atom is an electrical system having proper vibrations and that the periods of these vibrations give rise to the set of lines in the spectrum of an atom. In 1907 Conway made the bold supposition that each atom produces a single spectral line, so that the production of the spectrum is a collective effect of the atoms in a sample under investigation. This correct theory was put forward six years before the publication of the Bohr theory of the atom.

The first decades of the twentieth century were exciting for both theoretical and experimental physicists. In 1900 Planck proposed that energy is transferred discretely in integral multiples of the unit $h\nu$, where ν is the frequency of radiated energy and h is Planck's constant. Thus began quantum theory. Quantum theory and relativity provided new avenues of research for Conway.

In 1911 he showed that quaternions were especially well suited to express results in special relativity and to lead to new theorems. When later Dirac proposed his relativistic theory of the electron and Eddington his theory of protons and electrons, Conway expressed the four-by-four matrices employed in these theories very simply in quaternion form.

When the Bohr theory of the atom was extended by Sommerfeld and Wilson, Conway applied their quantum theory to show how the Zeeman triplet effect could be explained. He also showed how one could account for the ordinary spectrum of ionised helium and the s-term of orthohelium or parhelium, and could provide the correct value of the ionisation potential.

Immediately after the publication in 1926 by Uhlenbeck and Goudsmit of their paper on electron spin, Conway wrote a paper on the dynamics of a spinning electron. In this he provided a quantum mechanical theory in which the electron is regarded as a uniform rotating sphere that describes an orbit around a fixed nucleus. Then in 1927 he applied the de Broglie–Schrödinger wave mechanics, which he called 'undulating theory', to two-electron orbits and obtained the Rydberg form of series terms.

Conway was a man of genial disposition and with a boyish sense of humour. His university lectures were well ordered and were delivered with great speed. With the help of related textbooks one could build on his lectures to acquire a deep knowledge of the subject matter. When you heard him lecture on Hamiltonian mechanics or on some other topic to which he had made a personal contribution, he infused enthusiasm. He was admired and respected by his students and in particular by de Valera, who as taoiseach often sought his advice on academic matters.

Having spent his boyhood near the Wexford coast, Conway had a love of the sea and during most of his life he lived on the shore of Dublin Bay. He was a strong swimmer and enjoyed sailing. He was hospitable and, if you called unexpectedly during the evening, you could be sure of a warm welcome.

Further reading:
James Mc Connell (ed.), *Selected papers of Arthur William Conway* (Dublin, 1953).

James Mc Connell
School of Theoretical Physics
Dublin Institute for Advanced Studies

'Student' in 1908.

Born: Canterbury, 1876

Died: London, 16 October 1937

Family: He was the eldest of four sons and a daughter of Colonel Frederic Gosset, R. E., who married Agnes Sealy Vidal in 1875. The Gossets were an old Huguenot family.
  Married: Marjory Surtees Phillpot, youngest daughter of the headmaster of Bedford School, in 1906. She was captain of the English ladies hockey team and subsequently played for, and captained, the Ireland team.
  Children: One son Harry, who became a physician, and two daughters, Bertha and Ruth

Addresses:
Woodlands, Monkstown, Co. Dublin
Holly House, Newtownpark Avenue,
  Blackrock, Co. Dublin.

The Dublin brewery which, since 1759, has borne the name Guinness, traditionally brewed porter or stout from malted barley, flavoured with hops and darkened by roasted malt. In the latter part of the nineteenth century its product penetrated markets in Britain and overseas, and the brewery expanded enormously. This demanded a more scientific approach to quality control, and the board set about appointing persons with a scientific background as Brewers, the senior men in charge of the process. In 1899, W.S. Gosset, a scholar of Winchester and of New College, Oxford, and a graduate in mathematics and natural science, was appointed as a Brewer.

Gosset soon became aware of the need to be able to estimate experimental errors, the problem being that often only a small number of samples was available. It took at least a day to obtain the result of changing one variable in the experimental brewery, while in the development of new strains of barley, in which Guinness worked closely with the Department of Agriculture and its cereal station at Ballinacurra, the testing of one 'cross' might take a year. The usual method of repeating each experiment many times was therefore impracticable. In 1904, Gosset wrote a report on 'The application of the law of error to work of the Brewery', as a result of which he was put in touch with Karl Pearson (1857–1936), the mathematician who had been building up the large biometric laboratory at University College London and was probably the leading statistician of the day. Pearson had recently introduced the chi-squared goodness-of-fit criterion and founded the journal *Biometrika*.

Pearson's work was on large samples, but in 1906 Gosset spent two terms in London on the application of statistics to small samples, and it was during that period that he laid the foundation for the basis of his most famous paper, 'The probable error of a mean', published in *Biometrika* in 1908. (At that period, Guinness did not allow its staff to publish under their own names, and for this reason Gosset used the pseudonym 'Student' for all of his 22 published papers.)

In this paper, by examining by a mixture of theory and practice how means of small samples were distributed, he produced tables from which could be computed the probability that the population mean would lie within certain numbers of standard deviations of the sample mean for sizes of sample 4 to 10.

Ronald Aylmer Fisher (later Sir Ronald, F.R.S., 1890–1962), a mathematical student at Cambridge, wrote to Gosset giving a more rigorous proof of the distribution and recommending a change from n to n-1 in calculating the variance. When this was done, the statistic became known as Student's t (see box) and the t-test entered the language of statistics. Fisher became statistician to Rothamsted Experimental Station at Harpenden, and it was largely due to his advocacy that Student's t became widely accepted as a test of significance.

Gosset's interests extended to correlation co-efficents, randomisation of field trials (he had differences here with Fisher, though they remained friends), genetics, and other areas of statistics. However, it would be a mistake to think of him as a 'statistician employed by Guinness', or as a 'brewer with a sideline in statistics'. His brewery work made statistics a necessary tool; where the tool was inadequate he improved it. His was an essentially practical approach; he developed methods that worked, being content to leave mathematics to Fisher. On the other hand, he could test the practical usefulness of the mathematics.

In these electronic days, it is easy to forget that the calculations of the vast tables for z or its successor t had all to be done 'by hand' or on what would seem to us primitive mechanical calculators. Much of this was done by Gosset himself at home, though he had some assistance in the Brewery from W.A. Bowie and, from 1922, E. Somerfield.

A Use of Student's t

We wish to construct a 95% confidence interval estimate of the mean yield of barley per square metre μ on a farm. We take a random sample of size n plots each of 1 square metre area, find the yield x in each, and the sample mean $\bar{x}$. Were we to know the population variance σ^2 of the yields per square metre, then based on the normal probability tables $\bar{x} \pm (1.96\ \sigma/\sqrt{n})$ would with probability 0.95 contain the unknown mean μ. When σ^2 is not known, it is usually estimated by the sample variance

$$s^2 = \frac{\sum(x - \bar{x})^2}{n - 1}$$

For large n (above say 30), $\bar{x} \pm (1.96s/\sqrt{n})$ constitutes an approximate 95% confidence interval for μ. 'Student', however, showed that for small n such interval estimates were much less reliable than 95%, and that in such a situation 1.96 should be replaced by the appropriate t value with $n - 1$ degrees of freedom. Values of

$$t = \frac{(\bar{x} - \mu)\sqrt{n}}{s}$$

are obtained from the curves and tables for this function which he had calculated.

For example, if $n = 12$, $\bar{x} = 325$ g and $s^2 = 125$, the appropriate t value for 11 degrees of freedom is 2.201, and the resulting 95% confidence interval for the mean yield per square metre would be

$$\bar{x} \pm 2.201s/\sqrt{12} = [318, 332]$$

(I thank Dr Philip Boland, University College Dublin, for assistance.)

Gosset was a well-liked person, interested in the outdoor life, golf, gardening, sailing, skiing, walking and fishing; a little eccentric perhaps. He built several boats for fishing, one of which with a novel design with two rudders was featured in *The Field* in 1936.

In 1935 he left Dublin to take charge of the new Guinness brewery at Park Royal in London. He was able to supervise the start-up of this enterprise, but he died in 1937 at the early age of 61. His t-test and process of 'Studentisation' are his memorial.

Further reading:
E.S. Pearson and J. Wishart, *'Student's' collected papers* (University College, London, 1942).
C. Chatfield, *Statistics for technology* (1978).
D. Hoctor, *The Department's story: a history of the Department of Agriculture* (Dublin, 1971).
J.F. Box, *R.A. Fisher, the life of a scientist* (1978).

Adrian Somerfield
St Columba's College
Whitechurch
Dublin

Courtesy of the trustees of the Ulster Museum.

Born: Belfast, 15 December 1883

Died: County Down, 19 May 1972

Family: Son of James Stelfox, M.I.C.E.,
and Jennie McIlwaine
 Married: Margarita Dawson Mitchell 1914
 Children: Two sons and a daughter

Addresses:
1883–1914 Delamere, Chlorine Gardens,
      Belfast
1914–1920 Ballymagee, Bangor, Co. Down
1920–1956 14 Clareville Road, Harold's
      Cross, Dublin
1956–1972 21 Tullybrannigan Road,
      Newcastle, Co. Down

Distinctions:
Member of the Royal Irish Academy 1912
Honorary fellow of the Linnaean Society 1947
He refused two offers of honorary degrees.

Arthur Wilson Stelfox was educated in Campbell College, Belfast, and trained as an architect in London and Belfast. He obtained his associateship of the Royal Institution of British Architects in 1909, though he practised this profession little. He ran a market garden in County Down for a short period. In 1920 he was appointed to the Natural History Division of the National Museum of Ireland. For several years from 1924 he was acting keeper of the Natural History Division. However, he failed to be appointed to the post of keeper of the Division (on the grounds that he did not have a university degree!) in spite of his international standing in several fields of natural history and in spite of his achievements in the post in an acting capacity. Subsequently he was appointed deputy keeper. He retired in 1948 and later went to live in Newcastle, Co. Down.

He was a keen naturalist from a very early age, and his first publication dates from 1904. He became a leading authority in Britain and Ireland on non-marine molluscs, then on bees, wasps and ants, then on sawflies; he then began a vast study of the parasitic hymenoptera, such as the ichneumons and braconids. In addition to these main areas, Stelfox was a polymath in natural history: he was an expert in the identification of mammals and birds from bone fragments in cave deposits; he had a wide knowledge of flowering plants, particularly arctic–alpine plants and difficult groups such as sedges and willows; and he carried out genetical studies on several species of snail, involving long-extended breeding studies. He has been described as 'a naturalist of prodigious knowledge, intelligence, versatility and memory'. He used his gifts to stimulate and help many of the young naturalists of his time.

His biggest contribution was in the collecting and recording of insects and molluscs throughout Ireland and in many parts of Britain. He appreciated the importance of collecting long series of each species with detailed records on habitat. In the case of the parasitic hymenoptera, his collection contained over 100,000, perhaps 200,000, superbly mounted and catalogued specimens, many new to science: it is one of the greatest collections ever made by one person and has been described as a national treasure. Only a small fraction of the information contained in this collection has been published, and his collections and records will provide material for publication by generations of entomologists. It is a tragedy that the lack of government interest in the Natural History Division of the Museum, together with unfortunate official decisions affecting himself and some other people whom he respected, led to the conviction that his collections would never get the required degree of study in this country, and he donated all his collections and records outside the state. The parasitic hymenoptera collection went to the Smithsonian Institution in Washington. Happily, duplicate specimens from this collection have been returned recently to the National Museum, and these will enable his work to be carried on by entomologists in Ireland.

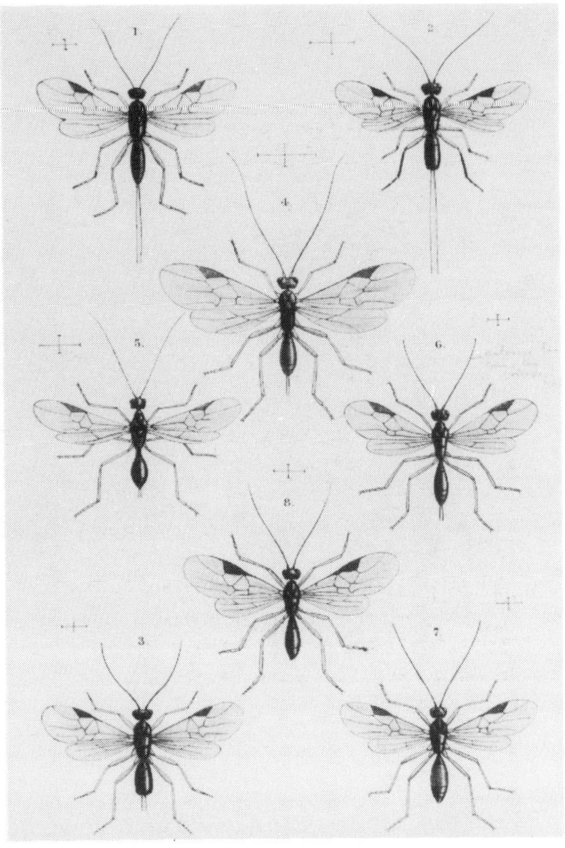

Plate X of T.A. Marshall's (1889) A monograph of British Braconidae (Transactions of the Entomological Society of London) *annotated by Stelfox. Courtesy of The National Museum of Ireland.*

Further reading:

Various authors, 'Arthur Wilson Stelfox, 1883–1972'. *The Irish Naturalists' Journal* **17** (1973), 286–307.

N.F. McMillan, A.E. Ellis and L.M.Cook, 'Arthur Wilson Stelfox, 1883–1972'. *Journal of Conchology* **27** (1972), 520–33.

B.P. Beirne, 'Irish entomology: the first hundred years'. *The Irish Naturalists' Journal, Special Entomological Supplement* (1985). This includes a critical review of Stelfox's character and contributions. Many who knew him regard some of it as unfair, and parts of it have been replied to by G.F.Mitchell and N.F. McMillan *(The Irish Naturalists' Journal* **22** (1986), 122).

Frank Winder
Department of Biochemistry
Trinity College
Dublin

Joseph Doyle, P.R.I.A., in conversation with Eamon de Valera in the Royal Irish Academy.

Born: Glasgow, 25 March 1891

Died: Dublin, 11 April 1974

Family:
Married: Elizabeth J. Leonard, a lecturer in the Department of Botany, University College Dublin 1919
Children: Three sons and one daughter. His sons are all medical doctors; the youngest, J. Stephen, is professor of medicine in the Royal College of Surgeons in Ireland. His daughter, Mary Helen (m. M.W. O'Reilly), a lecturer in the Department of Botany, U.C.D. from 1945 to 1962, obtained a Ph.D. on conifer wood structure in 1948. She died in 1987.

Distinctions:
Member (1927), on Council at various periods from 1931 onwards, secretary (1957–63), vice-president (1963–64), president (1964-66), vice president (1966–67) Royal Irish Academy
Boyle Medal, Royal Dublin Society 1942
President, Botany Section, British Association for the Advancement of Science Meeting in Dublin 1957
Vice-president, Royal Dublin Society 1968-74

Addresses:

1920–1924	20 Victoria Avenue, Donnybrook
1924–1926	40 Marlborough Road, Donnybrook
1926–1933	6 Achill Road, Drumcondra
1933–1956	69 Anglesea Road, Ballsbridge
1956–1966	6 Park Drive, Rathmines
1966–1972	14 St Kevin's Park, Dartry
1972–1974	Carrigower, Mount Anville Road

Joseph Doyle graduated in natural science at University College Dublin in 1911 and, after a period of graduate research in Hamburg, joined the staff of the Biology Department in 1913. In 1924 he was appointed professor of botany in succession to J. Bayley Butler (who was simultaneously appointed professor of zoology in succession to George Sigerson). Thereafter Doyle was responsible for organising an independent Botany Department and directing it until his retirement in 1961. Over the period 1916–72 he published a long series of research papers on the reproductive biology of conifers, which established his international scientific reputation.

Conifers are woody seed plants, almost all trees, of widespread distribution in northern and southern hemispheres. Their reproductive biology had, in general outline, been determined in the last century, but many details still remained obscure until recent times. In common with all living seed plants, the dominant vegetative plant produces spores, which develop into microscopic structures bearing sex organs: female eggs in ovules and male gametes in pollen grains. Only the pollen grains are released from the spore-bearing plant. They are borne in the air by wind currents to the retained female eggs, and, after fertilisation, a complex series of transformations occurs in the young embryo before a new spore-bearing plant is formed and released eventually as a seed. The study of conifer embryology engaged Joseph Doyle throughout his scientific career. It called for great patience in obtaining the various stages of embryo development: these could only be investigated by the preparation of large numbers of thin sections of material which had been chemically arrested and stained for observation of cellular development.

Pollination is an important pre-fertilisation stage in the life cycle of seed plants. Doyle and his students in the 1930s devoted attention to pollination mechanisms in conifers: the reception of air-borne pollen by the female ovule. In 1945 he summarised his own and pre-existing observations in a review paper 'Developmental lines in pollination mechanisms in the Coniferales' *(Scientific Proceedings of the Royal Dublin Society* **24**, 43–62), which is still considered to be the definitive statement in modern textbooks. In several living conifers the ovule exudes a fluid, which traps pollen: Doyle attempted to show, with reference to their fossil precursors, that this was an evolutionarily primitive condition that has been suppressed in many other living conifers.

The stages subsequent to pollination were, however, those that particularly engaged Doyle and several collaborators over many years. Making use of the outstanding collections of mature conifers at Glasnevin, Kilmacurragh, Powerscourt, Rostrevor and elsewhere, they described the embryogeny of a series of southern-hemisphere conifers—*Athrotaxis, Callitris, Fitzroya, Podocarpus, Saxegothaea.* This new information supplied an important perspective for existing knowledge on northern-hemisphere genera, especially *Pinus,* which had hitherto provided the staple textbook examples of conifer reproductive biology. Much of the success of this research depended on the technical skill of Doyle's collaborator W. S. Looby; they published seven papers together. Doyle soon came to the conclusion that some of these southern-hemisphere conifers, such as *Podocarpus andinus,* showed a simpler scheme of embryological development than that found in *Pinus.* A masterly summary of the evidence was presented in 'Pro-embryogeny in *Pinus* in relation to that in other conifers— a survey' *(Proceedings of the Royal Irish Academy* **62** (13), 181–216, 1963). It is a lasting legacy of his research that the great variability in embryological events found in conifers may now be seen more clearly as modifications of a relatively simple basal plan.

Doyle, however, did not commit himself so far as to conclude that such modifications represented an evolutionary series that was reflected in the systematic affinities of modern conifers: the embryological trends seemed to be largely independent of the systematic trends based on a variety of other morphological and anatomical criteria. This led him to question, notably in his address to the British Association meeting in Dublin in 1957, the concept of natural selection as an important evolutionary factor; he treated it as 'a hypothesis which is plausible enough in certain minor instances'.

Many conifers, during the course of development of a single fertilised egg, show a proliferation of embryos (often dozens); this phenomenon is known as 'cleavage polyembryony'. As only one of the embryos typically survives to maturity in the seed, the evolutionary significance of this process is obscure. By some it has been argued that it provides the female plant, on which the young embryo depends for nutrition, with an opportunity to select some embryos at the expense of others. But nothing is known of the existence, if any, of genetic variability between the cloned embryos, nor of its significance for the future growth of the successful embryo. To this topic, with the experience of over fifty years, Doyle turned in his last two-part paper with his colleague Martin Brennan, S.J.: 'Cleavage polyembryony in conifers and taxads — a survey' *(Scientific Proceedings of the Royal Dublin Society* A4 (6, 10), 1971/2). Here they considered the wide array of existing evidence, and concluded that the absence of cleavage of an embryo was the more primitive condition from which cleavage arose independently in various evolutionary lines of conifers. Once again *Pinus,* for long treated as typical of conifers embryologically, was considered to be a specialised and derived type, not a prototype.

Almost all of Doyle's research papers on conifer embryology, it may be noted, were published by the Royal Dublin Society, of which he was a devoted, life-long member, serving on many sub-committees over a long period. In the first half of this century the Society's *Scientific Proceedings* achieved international eminence botanically for three remarkable series of papers: those of Henry H. Dixon, F.R.S. and his collaborators in Trinity College Dublin on plant physiology, of Paul A. Murphy and colleagues in University College Dublin on plant pathology, and of Doyle and his collaborators on conifer embryology.

James White
Department of Botany
University College
Dublin

Born: Cork, September 1893

Died: Dublin, February 1973

Family: He came from a professional family; his father was city analyst, one brother, Stephen, was city engineer, another, Ben, was a classical professor and a well known Marxist.
  Married: Doreen Synge, who belonged to a widely flung family, well known in botany, literature and mathematics.
  Children: They had two sons and one daughter.

Addresses :

1920s	Pembroke Road, Dublin
1928–1932	19 Dawson Street, Dublin
1932–1964	Ticknock, Co. Dublin
1964–1973	Greenfields, Milltown, Dublin

Distinctions :
D.Sc. National University of Ireland
Honorary Sc.D. University of Dublin
President, Geographical Society of Ireland
  1943–46
President, An Taisce 1964–67

Trained at University College Cork as an engineer, Farrington joined the Irish Geological Survey in 1921 and soon found himself mapping in the area around Blessington, Co. Wicklow. Evidences of former ice-action, such as changes in drainage-pattern and deltas of an ice-dammed lake, were very prominent in this region and roused in him the interests in geomorphology and glaciology that he was to pursue for the rest of his life.

When a student in Cork, Farrington developed strongly nationalistic views, and he used to tell how visiting correspondents from English newspapers would be brought to interview him as a 'Protestant Sinn Feiner'. When he moved to Dublin he lived in a 'commune' with Diarmuid Coffey, who later became clerk of the senate, and Liam Price and Dorothy Stopford, who later married. Liam had been active in the republican courts, and later became a district justice. Tony Farrington married Doreen Synge, another member of the group; she was a close relative of J.M. Synge, the author.

In 1928 he changed step again, joining the Royal Irish Academy as resident secretary and librarian. He later also edited the *Proceedings.*

Together with Praeger and Mahr, Farrington organised the Quaternary Research Committee that brought Knud Jessen to Ireland in 1934 and 1935. Farrington worked with Jessen in the field, and played no small part in a dramatic advancement of knowledge of the glacial and vegetational history of Ireland. He was a founder member, editor and officer of the Geographical Society of Ireland, and a founder member and officer of An Taisce.

And all the time Farrington was continuing his own researches and producing a steady stream of papers, which won him both local and international recognition. He was the first worker in Ireland to make systematic counts of the stones in glacial deposits in order to categorise them and trace them from one area to another. It can fairly be said that for about thirty years, from 1935 to 1965, Farrington was Irish glacial geology.

```
                                                              1947    2

Wexford          Ardcavan          Clay at cliff base
Dry weight                          4lb       1oz
Grey-brown dense clay with light-brown partings -not stony
Dry weight after washing
                   ¼"              3⅜ oz
                  -¼"      1lb      2  oz .................. 28%
         Clay (by loss)   2lb     11¼ oz ................. 66%
                          ___________
                          4lb        1  oz
  -¼"
         Grey sand with pinkish tinge - clear quartz, well rounded grains
            - the small pebbles passing seive are mostly well-rounded - many
         shell fragments.
  +¼"
         Total number of pebbles          65
             A               33
             R₃              12
             R₂              11
             R₁               9

Very little igneous rock - several pieces of limestone - six pieces of
flint (five blank and one white) - one piece of chert - one piece of
O.R.S. congl.  Some of the limestone was of a pale brownish colour, not
at all like Carbs Lst.
This is a typical far-travelled boulder-clay
Sample brought in by G.F. Mitchell   January 1947      A.F.
```

A typical Farrington stone-count sheet.

Further reading:
See *Irish Geography*, volumes 1-11, *passim*.

Frank Mitchell
Department of Botany
Trinity College
Dublin

Born: Dublin, 11 April 1896

Died: Dublin, 8 February 1983

Family: The oldest child in the family of two sons and two daughters of Ned Geary, a Dublin civil servant, and Jennie O'Sullivan. Two of his cousins on his mother's side received honorary doctorates.
  Married: Mida Maura O'Brien 1927
  Children: Colm and Clodagh (Dooney)

Distinctions:
President, Statistical and Social Inquiry Society of Ireland 1946–50
Fellow, Econometric Society 1951; Council member 1962–4
President, International Statistical Institute
Honorary fellow, Royal Statistical Society 1957
Vice-president, Royal Irish Academy 1963–5
Honorary fellow, American Statistical Association
Chairman of Council, International Association for Research in Income and Wealth
Honorary doctorates National University of Ireland 1961, The Queen's University of Belfast 1968, Trinity College Dublin 1973
Boyle Medal of the Royal Dublin Society 1981

Addresses:
1896–1928   74 Upper Drumcondra Road, Dublin
1928–1936   3 Parnell Square, Dublin
1936–1966   27 Leeson Park, Dublin
1966–1983   12 Court Flats, Wilton Place, Dublin
(He lived in Paris 1919–21, Cambridge 1946–47, and New York 1957–60.)

The education of Ireland's greatest statistician, Roy Geary, began at the O'Connell Schools in Dublin and continued at University College Dublin, where he graduated with a first class B.Sc. in 1916, and at the Sorbonne in Paris, 1919–21. He spent most of his working life in Dublin — from 1923 to 1949 in the Statistics Branch of the Department of Industry and Commerce, from 1949 to 1957 as director of the newly established Central Statistics Office, and from 1960 to 1966 as director of the Economic Research Institute. He remained there (it is now the Economic and Social Research Institute), as consultant, from 1966 until his death.

His international reputation rests mainly on his contributions to mathematical statistics, although he also contributed to national accounting and headed the National Accounts Branch of the Statistical Office of the United Nations in New York from 1957 to 1960. His best known theoretical papers were published between 1925 and 1956, although he continued to publish theory much later into the 1960s and 1970s. While his work was almost always motivated by problems of practical importance, it contains many elegant and theoretically attractive results, e.g. that independence of mean and variance implies normality and that maximum likelihood estimators minimise the generalised variance for large samples.

A systematic analysis of his publications reveals three major identifiable though related strands. First, there was his continuing interest in the sampling theory of ratios and his 1930 derivation of the density of the ratio of two normal variates remains one of the few essential references in the field. The second major

Roy Geary with daughter Clodagh Dooney and grand-daughter Jill Dooney at Trinity College Dublin in 1973, when he was conferred with an honorary doctorate.

theme was testing for normality and enquiring into the robustness of inferential methods which depended formally on normality. This work is exemplified by his 1935 *Biometrika* paper where a suggested test is based on the ratio of the mean deviation to the standard deviation. The third major theme concerned the estimation of relationships where the variables are subject to errors of measurement. His influential 1949 paper has led to his being regarded, with Reiersol, as the leading pioneer of the now standard technique of instrumental variables, these being introduced by him in the context of errors in variables.

Later in life he turned more towards practical economics and the direct application of statistical methods. Perhaps this reflected his position as the first director in 1960 of the Economic Research Institute, founded to meet the need for more economic research in Ireland and to enlarge the knowledge of the economic and social conditions of Irish society. Among his first tasks as director, apart from administration, was a study of inter-industry relationships based on input-output data and a study of the Irish woollen industry. It is now generally accepted that he established a spirit of independent inquiry in the Institute from the beginning and that his international reputation was of great significance in building its initial successes and reputation.

While his zest for work never diminished — he wrote more than half of his 112 published papers after the age of 65 — he had many other interests. Among his life long loves were music, the theatre, reading, humour, soccer and children. He greatly admired France, regarding the French as the most civilised of nations. He longed for more rationality in both Irish and international politics and in public affairs. He carried over his high standards of work to all areas of his life and was unsparing of his time with young research workers, many of whom were greatly encouraged by him.

Further reading:

R. C. Geary, 'The frequency distribution of the quotient of two normal variates'. *Journal of the Royal Statistical Society* **43** (1930), 442–6.

R. C. Geary, 'The ratio of the mean deviation to the standard deviation as a test of normality'. *Biometrika* **27** (1935), 310–32.

R. C. Geary, 'Determination of linear relations between systematic parts of variables with errors of observation the variances of which are unknown'. *Econometrica* **17** (1949), 30–58.

J. E. Spencer, 'The scientific work of Robert Charles Geary'. *Economic and Social Review* **7** (1976), 233–41.

J. E. Spencer, 'Robert Charles Geary — an appreciation'. *Economic and Social Review* **14** (1983), 161–4.

J. R. N. Stone, 'Robert Charles Geary (1896–1983)'. In J. Eatwell, M. Milgate and P. Newman (eds), *The new Palgrave: a dictionary of economics, Vol. 2* (London 1987), p. 491.

John E. Spencer
Department of Economics
The Queen's University
Belfast

J. J. Nolan.

P. J. Nolan.

Addresses:
J. J. Nolan

Until 1941	Ros-na-Greine, 14 Avoca Avenue, Blackrock, Co. Dublin
1941–1945	30 Fitzwilliam Place, Dublin
1945–	Lugnaquilla, Cowper Road, Rathmines, Dublin

Addresses:
P. J. Nolan

Until 1937	65 St Stephen's Green, Dublin
1937–	35 Eglinton Road, Donnybrook, Dublin

In the early years of this century two remarkable brothers entered University College Dublin to study for a degree in physics. Both were later to become professors in the college and helped to establish the Atmospheric Physics Research Group which is still thriving today. John J. and Patrick J. Nolan were born in Omagh, Co. Tyrone, where they received their early education and were awarded scholarships to study in U.C.D.

J. J. Nolan (the elder brother) graduated from U.C.D. in 1909 and joined the staff of the university in the following year. In 1920 he succeeded J. A. McClelland as professor of experimental physics. He was appointed registrar of the college in 1940 and held both posts until his death in 1952.

He published his first paper with McClelland in 1911 on the subject of the electric charge on rain, which signalled the start of a lifelong interest in atmospheric physics in general and atmospheric electricity in particular. His research carried him into many aspects of this branch of physics research and in each case he made a significant contribution to the fund of knowledge on the particular topic. However, it was his work on the equilibrium of ionisation in the atmosphere that many consider to be his most important piece of research, much of which was done in collaboration with his younger brother.

P. J. Nolan entered U. C. D. in 1911 and graduated in 1914. The following year he was awarded the M.Sc. degree and won a travelling studentship. However, because of the war, he was unable to travel abroad and so the first two years of the studentship were spent in U.C.D. At the conclusion of the war he went to the Cavendish Laboratory in Cambridge and worked under Sir J. J. Thompson and Lord Rutherford. It is

Members of the Atmospheric Physics Research Group prior to a visit to Colwyn Bay.
L. to R. C. O'Brolchain (later professor of physics in U.C.G.), J. Hughes, chief technician in the Physics Dept, P. J. Nolan and J. J. Nolan.

Crowds out at Colwyn Bay.

interesting to note that amongst P. J. Nolan's fellow research students in the Cavendish at that time were E. V. Appleton, J. Chadwick, A. H. Compton and G. P. Thomson, all of whom subsequently were awarded the Nobel Prize for Physics.

After his period at the Cavendish P. J. Nolan returned to U. C. D. and began a long and very fruitful career as a lecturer and later as a professor in the physics department. Probably his greatest contribution in the field of atmospheric physics was his development, with L. W. Pollak, of the photoelectric nucleus counter that bears their names. This instrument, which is used to measure condensation nuclei, is now the standard instrument in use throughout the world. (Condensation nuclei are very small particles — about 10^{-8}m in radius — which are present in the atmosphere and on which water vapour condenses.)

Two incidents from the long career of these two distinguished physicists will show that scientific research can have its lighter moments. The first relates to 1929 when members of the group went to Colwyn Bay in North Wales to investigate the effects of a solar eclipse on atmospheric ions. After all the travel and preparation the expedition was literally a wash out because of heavy rain. Added to that, local interest caused crowds to gather at the site chosen by the Irish scientists. The ensuing pollution caused by cigarette smoke, car exhausts, etc. made measurements of atmospheric particles valueless.

The U.C.D. Atmospheric Physics Research Group at Colwyn Bay 1929:
L. to R. C. O'Brolchain, P.J. Nolan, J.J. Nolan and J. Hughes.

The second event occurred almost four years later when, early in 1933, Professor J. J. Nolan and his chief technician, Mr Jack Hughes, were making measurements on atmospheric ions and conductivity in a cottage in what was then a rather remote valley at Glencree in the Wicklow mountains. They went up to the cottage well prepared for a stay of some days. However, a heavy snowfall cut off communications. Some people in Dublin, fearful for their safety, convinced Mr Eamon de Valera, the new head of state, that they were in danger of death from hunger and exposure. He ordered the army out to rescue the professor and his technician. Time passed and having heard nothing, Mr de Valera, accompanied by his sons, Eamon and Vivian, set out to find out what was happening. At Enniskerry they found the army lorry which had returned there after getting stuck on the road up to the cottage. Mr de Valera and his sons pressed on and later reached the cottage. And, far from finding Professor Nolan and Mr Hughes suffering badly from the cold, found them instead in front of a roaring kitchen fire entertaining with a bottle of whiskey the two army officers who had made it up on foot.

Under the guidance of J. J. and P. J. the atmospheric research group gained in reputation and became internationally known as the 'Nolan School'. Together the Nolan brothers published over 80 scientific papers and it is a measure of the quality of their work that papers in this area of research being published today still quote the work of the Nolans, which in some cases had been published over fifty years ago.

Further reading:
'Professor J. J. Nolan', *Nature* **169** (1952), 1036.
'Professor J. J. Nolan', *Studies* **41** (1952), 317–22.
P. J. Nolan, 'The photoelectric nucleus counter'. *Scientific Proceedings of the Royal Dublin Society,* Series A, **4** (12) (1972), 161–80.

Tony Scott
Department of Physics
University College
Dublin

Born: Bristol, 4 October 1906

Died: Dublin, 16 July 1986

Family:
    Married: Monica Morrissey 1934
    Children: Four daughters — Barbara, Margaret, Síle and Edwina

Distinctions:
Member of the Royal Irish Academy 1942
Professor of experimental physics,
    University College Dublin 1953
Honorary D.Sc. Queen's University, Belfast 1962
Member of Council of Europe Committee
    on Higher Education 1956-1979
Member of Senate of N.U.I. 1959–77

Addresses:
1937-1952 Seskin, Bird Avenue, Clonskeagh, Dublin
1952-1986 Huntersmoon, Sydenham Villas,
           Dundrum

Professor Thomas E. Nevin was born in Bristol in 1906 and entered University College Dublin in 1924. After a brilliant academic career he did his early post-graduate work under Professor J. J. Nolan on the diffusion and mobility of ions in the atmosphere. On the basis of this research he was awarded an 1851 Exhibition Scholarship in 1929 and went to Imperial College of Science and Technology, London, where he studied molecular spectroscopy under Professor A. Fowler, F.R.S. Rejecting invitations to appointments in several English universities, Nevin returned to UCD as an assistant in 1931. He refurbished the spectroscopy laboratory, building, almost entirely with his own hands, the 21ft Eagle spectrograph which was to provide such excellent service in succeeding years. Here was accomplished his classical work on molecular spectra, in particular the analyses of the quartet transitions of O_2^+ and the septet transitions of MnH; the latter remains to the present day the most complex molecular spectra analysed.

In 1944 Nevin, although never losing his interest in the quantum interpretation of molecular structure, transferred most of his attention to the burgeoning field of cosmic rays and meson physics. He built a cloud chamber, for which he devised an ingenious electronic control circuit, and with it studied the interactions of mesons in lead. This work led to a fruitful collaboration with the School of Cosmic Physics (see p. 100). When the cloud chamber was superseded by the use of nuclear emulsions, Nevin was one of the first to adopt the new techniques. Together with Professor C. Ó Ceallaigh he established collaborative links with C. F. Powell at the University of Bristol; this in turn led to numerous and fruitful collaborations between the Dublin group and other European Centres, including CERN.

Nevin was appointed professor of experimental physics in 1953. Dedicated to UCD, he served on the Governing Body and was acting president when Dr J. Hogan retired. As dean he worked for the growth of a vigorous and dynamic science faculty in the new science building in Belfield. He had a remarkably wide knowledge of physics and always kept abreast of important new developments. As a result his teaching was wide-ranging and authoritative. His personal research accomplishments, although relatively limited, were of the highest quality. Perhaps his greatest achievements were the fostering of physics and science generally in Ireland, his initiation of new fields of research and the support he gave to others.

P.K. Carroll
Department of Physics
University College
Dublin

Trade card for Yeates & Son 1834–39.
Reproduced from the Trade Card Collection in
the Science Museum Library by permission of
the Trustees of the Science Museum, London.

The Dublin-based Yeates family business of scientific instrument makers survived for well over a century. Although members of the Yeates family appeared in street directories from 1769 variously described as weavers, drapers and shoemakers, later there were brass founders, cutlers and surgical instrument makers. Samuel Yeates, the first to be designated 'optician', when the word meant 'instrument maker' rather than 'spectacle maker', first appeared in the directory for 1790. He moved the shop to 2 Grafton Street in 1827, and after several changes of name — both George Yeates and Stephen M. Yeates were connected with the business on these premises — the firm became Yeates & Son in 1865. This name still survives in neighbouring Grafton Arcade, but the business there has had no connection with the Yeates family since the 1940s. Other members of the family opened their own scientific instrument businesses for short periods in Dublin during the nineteenth century, amongst them Kendrick, William, Thomas and Horatio.

Yeates & Son made and sold a wide range of scientific instruments, and many still survive to-day. Much of their material was aimed at the growing mid-nineteenth-century market for educational material: with an increased demand for scientific apparatus in schools and colleges, there was scope for supplying reasonably priced equipment locally. From 1852 until 1910 Yeates & Son advertised as 'Instrument Makers to the University', which implies that Trinity College was a good customer. (Instruments with the firm's signature survive in the Physical Laboratory and the School of Engineering. Other institutions with either a teaching or demonstration function at this time are finding that Yeates instruments have survived until today.)

Various members of the Yeates family were of an inventive frame of mind. One of Samuel's sons, George, published a paper containing his ideas on minor improvements to the theodolite in 1845; another, Andrew, wrote about the bearings of transit instruments. Andrew went to England in 1821 to work with the pre-eminent instrument-maker Edward Troughton then installing new apparatus at Greenwich Observatory. He eventually ran his own business at 12 Brighton Place, New Kent Road, London, between 1837 and 1873. His 'improved portable theodolite' was, however, offered for sale in the Dublin firm's 1887 catalogue, so that he must have remained on good terms with the family.

His brother George, back in Dublin, also produced papers on meteorology, while his son Stephen Mitchell Yeates, who seems to have run Yeates & Son from 1865 until his death in 1901, published a work on the barometer. Stephen became involved in some of the earliest experiments with the telephone in Ireland. Almost ten years before Alexander Graham Bell produced his system, a German, Phillip Reis, had produced a form of transmitter, and in 1865 Stephen Yeates demonstrated his modified version of this before the Dublin Philosophical Society 'when both singing and the distinct articulation of several words were heard through it, and the differences between the speakers' voices clearly heard'.

S.M. Yeates' improved form of G.J. Stoney's local heliostat, by Yeates & Son, Dublin c.1880. Courtesy, Trustees of the National Museums of Scotland.

Despite a little knowledge about some of the members of the family who worked in the firm, there are a number of unanswered questions about the structure of the business — just how much they made themselves, how much they bought in and retailed, whether they bought in parts and then assembled them, how many people were employed, and whether they made comfortable profits or found that competition with English firms was very fierce. It would appear that in the early part of the nineteenth century much of what they sold was made by them. This is borne out in mid-century by their success at trade exhibitions, where exhibitors could only show what they had actually manufactured. At the Great Exhibition of 1851 in London, Yeates & Son obtained an Honourable Mention for their surveying instruments. In 1862, at the London International Exhibition, they exhibited a large public barometer with a three foot diameter dial, designed by them in 1858 for use in agricultural areas and fishing villages where they claimed it was found 'extremely useful'. Besides exhibiting in London, they also had stands at the Cork Exhibition in 1852 and the Dublin Exhibition in 1853. By 1887 they were able to proclaim proudly: 'Six first-class silver medals and other prizes have been awarded to Yeates & Son by the councils of the various exhibitions, both at home and abroad, for the excellence of workmanship and construction of the instruments exhibited by them. Silver medal, Inventions Exhibition, London, 1885'.

In the 1880s Yeates & Son produced a series of four trade catalogues from which some information may be gleaned. Firstly, the range of material covered was extensive: electrical apparatus, optical instruments, heat apparatus and drawing instruments, surveying and general engineering instruments. Secondly, they admitted to very few items not being of their own manufacture, yet some, such as Gramme's dynamo-magnetic machine, were clearly imported. Thirdly, it is impressive that so many pieces were described as 'S.M. Yeates Improved' indicating that the firm was able to produce its own version of those items. Fifty years later an observer recalled: 'Some who, like . . . Yeates, manufactured a number of various instruments have found it impracticable to do so today. Yeates made the last big attempt about 1890, but they were forced to market most of their products through the medium of English firms and they lost money. The home market is too small and because of the advance of science the variety is too great. A world market is needed, and this requires not merely highly educated scientific brains, but also large organising ability and great capital resources.'

Apparatus by Yeates & Son may be seen in St Patrick's College, Maynooth, Trinity College Dublin and University College Dublin.

Further reading:
T. H. Mason, 'Dublin opticians and instrument makers'. *Dublin Historical Record* **6** (4) (1944), 133–49.
A. D. Morrison-Low, 'The trade in scientific instruments in Dublin, 1830-1921'. In J.E. Burnett and A.D. Morrison-Low, *'Vulgar and mechanik'* — *the scientific instrument trade in Ireland, 1650-1921* (Dublin and Edinburgh, 1989).
C . Mollan, *Irish national inventory of historic scientific instruments, interim report 1990* (Dublin, 1990).

A.D. Morrison-Low
Royal Museum of Scotland
Edinburgh

Armagh Observatory May 1983.

Towards the end of the eighteenth century, the then archbishop of Armagh, Primate Richard Robinson, moved his ecclesiastical capital to Armagh and proceeded to construct a number of fine buildings in the manner of the day. Conscious of the former importance of the city as an educational centre, he proposed to establish a university there. The Archbishop's death, in 1794, put an end to this proposal. However, the Observatory and the Old Public Library, which probably were to have been incorporated in the university, were then already in existence and remain to this day.

The Observatory's foundation in 1790 was by an act of parliament signed by King George III, himself an 'amateur' astronomer of considerable repute. The financial means for the Observatory's operation were provided for by the archbishop's generous endowment of lands in the diocese of Armagh. The ownership was vested in a Board of Governors made up chiefly of members of the Chapter of Armagh Cathedral with the primate as its chairman.

While this arrangement provided the first directors with the means to devote all their time to the study of the heavens, the financial outlook became much bleaker as a result of the Disestablishment of the Church of Ireland in 1869. With its income continuously shrinking, the Observatory had to be run on very low means until, after repeated applications to the government, more satisfactory arrangements were made in the 1930s. At present the Observatory's operation and maintenance is funded by the Department of Education in Northern Ireland through an annual grant-in-aid. The U.K. Science and Engineering Research Council (SERC) also supports certain aspects of the operation.

During the two hundred years of its existence, the contributions of the Armagh Observatory to science have been much greater than its size and the funds spent on its maintenance would suggest. This has largely been due to the energetic and productive directors and staff it has had. In the nineteenth century Dr Thomas Romney Robinson, the third director, was a well-known astronomer and meteorologist of his day. He was closely associated with many of the technical improvements in telescope design which were implemented by the earl of Rosse in Birr, Co. Offaly (Vol. 1, p. 84), and by the firm of Grubb, the Dublin telescope makers (Vol. 1, p. 60 and this Vol., p. 18). His interests were varied and he made many other important innovations including the famous Robinson Cup Anemometer to measure the velocity of wind, an early example of which still graces the roof of the Armagh Observatory. Climatological observations had already started in the Observatory's early days, but under Robinson their quality was enhanced. As a result of his efforts and of those who have kept this activity going ever since, the Observatory is now a Climatological Reference Station in the World Meteorological Organisation's world-wide network.

Rev. Dr Thomas Romney Robinson (1792–1882), third director of Armagh Observatory.

Dr John Louis Emil Dreyer (1852–1926), fourth director of Armagh Observatory.

Robinson was succeeded by J. L. E. Dreyer who, while at Armagh, produced his *New general catalogue of nebulae and clusters of stars* still widely consulted by astronomers all over the world today. Dreyer was also renowned for his historical studies, in particular those relating to the history of our understanding of the planetary system, and about the great Danish astronomer of the sixteenth century, Tycho Brahe.

The Rev. W.F.A. Ellison kept the Observatory ticking over in the period between the two world wars. His main contributions were on physical observations of the planets, especially Mars, and on instrumentation. He wrote a book, *The amateur's telescope,* and contributed substantially to the famous book *Amateur telescope making.*

Dr E.M. Lindsay took up the director's post just when the Observatory's finances were being given a surer basis. However, because of the second world war, new developments were delayed. When finally the world war was over, the Observatory evolved quite rapidly into a modern astrophysical institute. Lindsay recognised the great, virtually untapped potential of astronomical research in the southern hemisphere. He was instrumental in setting up the so called Armagh-Dunsink-Harvard (ADH) Telescope at the Boyden Observatory near Bloemfontein in South Africa in 1950. This was going to be the largest Schmidt telescope in the southern hemisphere for the next 25 years. As a collaborative venture between the two governments of the UK and Ireland and Harvard University it was a forerunner of the large collaborative observatories operating in the southern hemisphere today.

Of course, in order to be able to deal properly with the wealth of observational material from the Boyden Observatory, measuring instruments were needed at Armagh. These were obtained with the help of outside grants, as was also a new building, the first addition to the Observatory accommodation since the late nineteenth century. Variable stars in the southern, central parts of our Milky Way and in the Milky Way's nearest neighbour galaxies, the Magellanic Clouds, were the main objects of study.

At this time, also, an outstanding Estonian refugee astronomer, Dr Ernst J. Öpik (1893–1985), came to Armagh. He had been one of Lindsay's teachers at Harvard, but in 1947 was in exile in West Germany where Lindsay contacted him and arranged for him to join the staff at Armagh. Öpik's interests were wide-ranging. He started with publications, at the age of 19, about Mars and about the Perseid meteors. From that time on the flow of articles never stopped until his retirement at age 88. His scientific work ranged

Rev. Dr William Frederick Archdall Ellison (1864–1936), sixth director of Armagh Observatory.

Dr Eric Mervyn Lindsay (1907–1974), seventh director of Armagh Observatory.

from meteor astronomy, to planetary physics, stellar statistics, the theory of the internal constitution and energy generation of the stars, stellar photometry, and the intergalactic distance scale. He also wrote extensively about the ice ages, extraterrestrial life, and the scientific exploration of space.

Öpik made several scientific predictions that were later confirmed. These included the presence of craters on the terrestrial planets Mars and Mercury, and the existence of a comet cloud at the very edge of the solar system. As one of the founders of *The Irish Astronomical Journal,* and its editor from 1950 to 1981, he wrote many articles for the *Journal.* Of special fame were his shorter contributions in the 'News and Comments' section, where he offered his very personal, very incisive and very valuable views on many current astronomical affairs.

In 1974, when Dr Lindsay died, the Observatory had grown from a one-man enterprise before the war into a widely known and respected institute of astronomical research. Its funding had been given a sound basis and it had a dedicated staff of five scientists, a graduate student and four administrative and technical support staff. During the years since 1976, when Dr M. de Groot took up the directorship, the emphasis has been on keeping the Observatory at the forefront of astronomical research. It has engaged in several projects involving scientific satellites and first-class ground-based telescopes in many parts of the world.

For the reduction and analysis of this constant stream of new observational data, computing equipment was introduced at the Observatory. After an initial period using computers at The Queen's University of Belfast (QUB) through a telephone line, a mini-computer was purchased in 1979. Soon there was a dedicated data line to QUB, and an increasing number of ever more sophisticated terminals at the Observatory. The latest breakthrough is the purchase, with a grant from the SERC, of a microVAX computer which, together with further peripherals and a similar outfit at QUB, is now forming the Northern Ireland node of the United Kingdom's astronomical computer network STARLINK.

Astronomical research at Armagh today concentrates on the study of cool stars, which have very active atmospheres where complicated magnetohydrodynamic processes release enormous amounts of energy in very short times. These studies are done at many wavelengths from X-rays to radio waves, and involve a wide international network of collaborators. In parallel, solar activity and the radiative properties of

Dr Ernst Julius Öpik (1893–1985).

atoms and ions in hot plasmas are studied in collaboration with colleagues at QUB and elsewhere. Hot supergiants, which are among the brightest objects in the universe, and a number of peculiar stars complete the spectrum of modern astrophysical research at an institution which is surprisingly youthful for its nearly 200-year existence.

Directors of the Armagh Observatory:

1790–1815	The Rev. Dr James Archibald Hamilton (1748–1815)
1815–1823	The Rev. Dr William Davenport (d. 1823)
1823–1882	The Rev. Dr Thomas Romney Robinson (1792–1882)
1882–1916	Dr John Louis Emil Dreyer (1852–1926)
1917	Joseph Alfred Hardcastle (1868–1917)
1918–1936	The Rev William Frederick Archdall Ellison (1864–1936)
1937–1974	Dr Eric Mervyn Lindsay (1907–1974)
1976–	Dr Martin Jan Hugo de Groot (1938–)

Further reading:

Patrick Moore, *Armagh Observatory,* 1790–1967 (Armagh, 1967).
Dolores Crowe, 'Thomas Romney Robinson (1792–1882), Director of Armagh Observatory (1823–1882)'. *The Irish Astronomical Journal* **10** (1971), 93–101.
The Irish Astronomical Journal **10** (1972), Special Issue.
The Irish Astronomical Journal **12**, No. 3/4 (1975), Memorial Issue.
The Irish Astronomical Journal **17** (1986), 411–42.
John Butler and Michael Hoskin, 'The archives of Armagh Observatory'. *Journal for the History of Astronomy* **18** (1987), 295–307.

Mart de Groot
Armagh Observatory
Armagh

Bartholomew Lloyd, president 1835: 'The likeness is tolerable but the expression far too sickly and sentimental for the late provost'. (Evening Mail, 4 July 1838).

The British Association for the Advancement of Science was established in 1831 at York, a centrical city of the three kingdoms. Its announced aims were to promote scientific research, to encourage contact between scientists and to obtain greater public attention to science and support for it. In pursuing these aims and in avoiding direct competition with other institutions the Association has long adopted distinctive features.

It has always been a peripatetic body, holding its annual meetings in the provinces for just one week in the year. This mobility was central to the Association simply because it was British and not just English, Irish, Scottish or Welsh. Having been founded in York in 1831, the Association visited in 1832 and 1833 the two old English universities, Oxford and Cambridge. In 1834 and 1835 it proceeded to Edinburgh and Dublin, the Scottish and Irish capitals. After the success of the 1835 meeting, the Association has returned on ten occasions to Ireland: in 1843 to Cork, 1852 Belfast, 1857 Dublin, 1874 Belfast, 1878 Dublin, 1902 Belfast, 1908 Dublin, 1952 Belfast, 1957 Dublin and Belfast 1987. With one exception the Irish meetings have been held in the two big cities, Dublin and Belfast, which around 1900 had populations of about 400,000 and 350,000 respectively. These two cities were large enough to guarantee adequate local support, to provide organisational prowess and to offer adequate accommodation in public buildings for the various scientific sections of the Association, which met simultaneously. Only large towns could cope with the sudden wave of visitors who stayed either in hotels or in private houses. Belfast and Dublin had the essential advantage of having universities whose academics played twin roles: local activists who carried out the ground work for the meeting in its many facets, and local savants who acted as officials in the various scientific sections.

By the mid 1830s the Association had begun to give research grants to selected individuals and projects and to achieve success as a scientific pressure-group on the British government. For both functions large attendances were needed to generate healthy receipts from the low membership fees and to enable the Association, by weight of numbers attending, to claim justly that as the 'Parliament of Science' it spoke for the British body-scientific. The Association rarely adopted the role of a missionary trying to stir up a few members of latent scientific warmth in the provinces. It had tried to do so in 1843 at Cork and unwittingly became embroiled in the political agitation about repeal of the Act of Union with England. The town of Cork warmly supported Daniel O'Connell's campaign, but the Tory gentry of the surrounding country were bitterly opposed to repeal. This local political division ensured a low attendance at the Cork meeting and poor receipts, so that half of the total sum granted for research had to be paid from the Association's capital. For the geologist Murchison, then the Association's general secretary, the Cork meeting was a disaster akin to being shipwrecked. Subsequently the Association stuck for decades to a small circuit of major manufacturing and commercial cities and of university towns. About 1900 big English and Welsh cities which had acquired university colleges were inserted into the circuit; but the Irish gyro was never extended beyond Dublin and Belfast. Cork was not forgiven for the fiasco of 1843 so the Association has not returned to it; and places such as Galway and Limerick have never been visited.

Local pride has often been stimulated by the touring scientific gala. Local members of the Association were keen to show that their locality was not lagging behind in the march of intellect and civilisation; they often went out of their way to promote their area and themselves. In 1835 Philip Dixon Hardy, a Dublin printer who was the first in Ireland to use steam printing, decided to capitalise on the scientific frenzy aroused by the Association: he published as a commercial venture a volume detailing the proceedings of the meeting. Such local publications helped to consolidate a favourable local public image of science. Entrepreneurship and local pride were nicely shown in 1874 when the Belfast Naturalists' Field Club prepared a *Guide* for the members of the Association informing them about the local features of scientific interest. Subsequently other places followed the lead given by Belfast and produced a series of authoritative local *Handbooks.*

Hospitality and ceremony have long been hallmarks of the Association, nowhere more so than at Dublin. This tradition began in 1835 when Trinity College conferred its highest distinction, the honorary LL.D., on five visiting

Humphrey Lloyd, son of Bartholomew Lloyd , president 1857.

savants. On the final day, the fellows of Trinity and Bartholomew Lloyd, provost of Trinity and president of the Association, gave a farewell banquet in the Examination Hall to three hundred of the Association's members. Before the dinner the young Irish mathematical prodigy, William Rowan Hamilton, was ceremoniously knighted in front of the assembled brotherhood of science. After the feast the members were rendered *hors de combat* by indigestion. Turtle was such a staple item on all the Dublin menus that one wag neatly described the peripatetic philosophers as 'gastropatetics' who met primarily to discuss the comparative anatomy and gastronomic phenomena of *Chelonia migdas*, called by the unassociated, turtle.

The Association's festive hospitality and spectacle helped to make science visible and into news. Its annual assemblies usually attracted huge and peaceful audiences, mainly indoors but sometimes *al fresco.* The Association tried to proclaim the difference between scientific calm and party factionalism: scientific fêtes in Phoenix Park, Dublin, were welcome relief from Ireland's political turmoil. This concern with presenting science as neutral ground and as a soothing balm has always characterised the Irish meetings. It accounts for the two visits, one to Northern Ireland and the other to the Republic, made in the 1950s and explains why the economics section was able to act as mediator between the contending parties in the linen mills strike in 1874 at Belfast and to bring it to a speedy end.

If in general the Association has tried to soothe inflammation caused by political and religious controversy, it was not entirely successful in 1874. Like most of his predecessors (Bartholomew Lloyd 1835; the third earl of Rosse, 1843 (Vol. 1, p. 85); Humphrey Lloyd, 1857 (Vol. 1, p. 32)) but unlike his successors, the president that year was an Irishman, the Carlow born John Tyndall. The Lloyds, both of whom were clergymen, wished to demarcate science from religion in a way which, while guarding against materialism and atheism, preserved the essential religious truth of scripture. They proclaimed

General meeting in the Corn Exchange, Cork 1843 (Illustrated London News).

that science should be pursued with humble reverence and religious awe. In contrast, Tyndall's 1874 presidential address seemed to advocate a reductionist attitude to science and not to oppose materialism and atheism. He also claimed that scientific knowledge represented the highest level of human experience and referred to science and theology as adversaries. At the end of his address the atmosphere began metaphorically to smell of brimstone. Tyndall and Huxley, with a lecture on animal automata, outraged local Presbyterians, who denounced the blasphemous atheism of the 'atomic, molecular chiefs'.

The British Association has been hailed as the Philosophical Carnival, as Vanity Fair, as the Philosophical Olympiad, as Frankenstein's monster, and as the British Ass. These various sobriquets show that the Association has provoked a wide variety of responses just as it has continued to provide varying opportunities and types of participation for its members. In the case of Irish meetings, different individuals have used the Association to serve their own needs. Three examples must suffice. In 1843 Joule bothered to travel from Manchester to Cork, where he revealed his first ideas about the mechanical equivalent of heat to a silently disapproving chemistry section. In 1852, the young and ambitious Huxley, desperate for a job, journeyed from London to Belfast in order to impress the assembled Irish scientists. Others had noticed that women were present in force at the Association's social events. A critic of the 1835 Dublin meeting stressed that with the aid of concerts and balls, beautiful women, sound claret and strong whisky, the sages made out remarkably well. One botanical sage, Joseph Dalton Hooker, even met both his wives at meetings of the Association. In 1835 the ageing Irish poet Thomas Moore was less ambitious: he merely prowled *à la Byron* at a soirée in search of pretty girls. The success of the Association, in Ireland as elsewhere, has indeed depended on its ability to serve the varying interests of its members by developing a wide range of accessible functions, latent as well as manifest.

The framed facsimile reads:

A D D R E S S

BY

PROFESSOR SIR WILLIAM R. HAMILTON.

IT has fallen to my lot, Gentlemen, as one of your Secretaries for the year, to address you on the present occasion. The duty would, indeed, have been much better discharged had it been undertaken by my brother secretary; but so many other duties of our secretaryship had been performed almost entirely by him, that I could not refuse to attempt the execution of this particular office, though conscious of its difficulty and its importance. For if we may regard it as a thing established now by precedent and custom that an annual address should be delivered, it is not, therefore, yet, and I trust that it will never be, an office of mere cold routine, a filling up of a vacant hour, on the ground that the hour must be some way or other got rid of. You have not left your homes—you have not adjourned from your several and special businesses—you have not gathered here, to have your time thus frittered away in an idle and unmeaning ceremonial. There ought to be, and there is, a reason that some such thing should be done; that from year to year, at every successive reassembling, an officer of your body should lay before you such an address; and in remembering what this reason is, we shall be reminded also of the spirit in which the duty should be performed. The reason is the fitness and almost the necessity of providing, so far as an address can provide, for the permanence and progression of the body, by informing the new members, and reminding the old, of the objects and nature of the Association, or by giving utterance to at least a few of those reflections which at such a season present themselves respecting its progress and its prospects; and it is a valid reason, and deserves to be acted upon now, however little may have been left unsaid in the addresses of my predecessors in this office. For if even amongst the members who have attended former meetings, and have heard those eloquent addresses delivered by former secretaries, it is possible that

1835.   d

Start of the address by William Rowan Hamilton (Vol. 1, p. 36) to the 1835 Dublin meeting.

Further reading:

P. D. Hardy, *Proceedings of the fifth meeting of the British Association for the Advancement of Science held in Dublin, during the week from 10th to the 15th August, 1835, inclusive. With an alphabetical list of members enrolled in Dublin* (Dublin, 1835).

O.J.R. Howarth, *The British Association for the Advancement of Science: a retrospect 1831–1931* (London, 1931).

R. MacLeod and P. Collins (eds), *The Parliament of Science: The British Association for the Advancement of Science 1831–1981* (Northwood, Middlesex, 1981).

J. Morrell and A. Thackray, *Gentlemen of Science: early years of the British Association for the Advancement of Science* (Oxford, 1981).

Jack Morrell
School of European Studies
University of Bradford

Ireland holds a position of esteem in the annals of ninteenth-century medical history for the remarkable contributions to clinical medicine emanating from a movement that came to be known simply as the 'Dublin School'. This title, correct in denoting the origins of the school and its role in the reformation of clinical medicine, does not, however, convey the dynamic idealism and iconoclasm that gave to this renaissance international recognition and the approbation of posterity. The school has been decked with many garlands, not least of which is the romantic title 'the golden age of Irish medicine', a tribute not undeserved, for at no time previously, nor at any time since, has Dublin had so great an influence in medicine.

Three giants stand out from a galaxy of lesser, though by no means insignificant, luminaries who constituted the 'School' — Robert Graves, Dominic Corrigan and William Stokes (Vol. I, p.34). If we seek qualities common to these three Irishmen, we may discern two outstanding talents: a compelling desire to observe the pattern and effect of illness with impartiality even when their studies refuted conventional practice, and the ability to describe their observations with elegance and authority. The personalities of the founders may fade into the shadows of time, but their contributions to medicine have been immortalised by eponyms with which medical students across the globe are familiar — 'Graves' disease', 'Cheyne–Stokes respirations', 'Stokes–Adams attacks', 'Corrigan's disease', and 'Corrigan's pulse'.

There were many doctors in the 'Dublin School' whose lives followed a less hectic course than those of the three founders and whose contributions, though more modest, were nonetheless significant. Among these were the ophthalmologist Arthur Jacob (1790–1874), Robert MacDonnell (b. 1828), who first performed surgery under anaesthesia in 1847, Francis Rynd (1801–61), inventor of the hypodermic syringe, which allowed doctors for the first time to give morphine by injection rather than by mouth for the relief of pain, and that towering example of the Victorian polymath, the Rev. Dr Samuel Haughton (see p. 36), divine, scientist and physician, who is possibly best known for 'Haughton's drop', a calculation giving the length of the drop needed to dislocate the cervical spine and so cause instantaneous death in hanging, rather than slow strangulation.

To many it may come as a surprise to find William Wilde (1815–76) included as a member of the 'Dublin School', but then unfortunately this great Victorian is often remembered only as the father of Oscar, or he is ridiculed and lampooned for his eccentricities and illicit amours. Too often it is forgotten that he was an innovative doctor, an accomplished archaeologist and author of some very fine books on Ireland. In 1844, he opened St Mark's Ophthalmic Hospital and Dispensary for Diseases of the Eye and Ear in Mark Street, and for many years this was the only hospital in the British Isles teaching both aural surgery and ophthalmology.

The 'Dublin School' began about 1830 and lasted scarcely fifty years. Its success was dependent foremost on the extraordinary energies and talents of its main progenitors, Graves, Stokes and Corrigan. Others of ability were to follow but they failed to sustain the spirit of the 'School'. We may well wonder why so vibrant a movement was permitted to decay. The conditions in which subsequent generations practised were not substantially different from those of the mid nineteenth century; there were the same hospitals with the addition of some new ones; there were more doctors and nursing improved greatly; a limited amount of money for research became available whereas there had been no provision for research funding in Victorian Ireland; the government participated in health care, not always acting in the best interests of the sick but, nonetheless, augmenting greatly the voluntary support on which mid-nineteenth-century medicine depended. And yet the School disappeared. The *raison d'être* of the 'Dublin School' was its iconoclasm, which was fuelled from without rather than within Ireland. The members of the School competed with and enjoyed the company of the European leaders of medicine; their ideals and their standards were pitched well above the mediocrity which Ireland, through complacency, an insular philosophy, and often unawareness of anything better, is prepared to tolerate. Had later generations been prepared to seek and absorb the influence of European and American medicine, the school might have survived, and Irish medicine might have been saved from a period of stagnation and apathy from which it only now shows some feeble signs of emerging. If today's medical profession is to be enriched from a study of the rise and rapid decline of the 'Dublin School', it will be by the realisation that its future lies not

Robert James Graves. Sculpture in marble by Albert Bruce Joy. By courtesy Royal College of Physicians of Ireland. Photograph by D. Davison.

Dominic John Corrigan. Sculpture in marble by John Foley. By courtesy Royal College of Physicians of Ireland. Photograph by D. Davison.

within the narrow confines of the island that is Ireland, but beyond in the broader intellectualism of international science.

Further reading:

E. O'Brien, *Conscience and conflict—a biography of Sir Dominic Corrigan, 1802–1880* (Dublin, 1983).

E. O'Brien, A. Crookshank and G. Wolstenholme, *A portrait of Irish medicine — an illustrated history of medicine in Ireland* (Dublin, 1984).

W. Stokes, 'The life and labours of Graves'. In W. Stokes (ed.) *Studies in physiology and medicine by the late Robert James Graves* (London, 1863), 9–83.

J. F. Duncan, 'The life and labours of Robert James Graves'. *Dublin Journal of Medical Science* **65** (1878), 1–12.

S. Taylor, *Robert Graves. The golden years of Irish medicine* (London, 1989).

ROBERT JAMES GRAVES   *Physician*   1796–1853

Born: Dublin 1796

Died: Dublin 1853

Family: Son of Richard Graves, scholar and divine, who was twice Donellan lecturer, Archbishop King's professor of divinity, professor of laws and regius professor of Greek and divinity at Trinity College, and dean of Ardagh.

Distinctions:
King's professor, Trinity College Dublin 1827
President of the Royal College of Physicians
   of Ireland 1843
Fellow of the Royal Society London 1849
Honorary member of the medical societies
   of Berlin, Vienna, Hamburg, Tübingen,
   Bruges and Montreal

Address:
Merrion Square, Dublin

The international reputation of the 'Dublin School' can fairly be stated to have had its foundations in Robert Graves. He was born in Dublin in 1796 to a family whose ancestors had come to Ireland with the Cromwellian army. Having spent some time at Edinburgh, Graves graduated from Trinity in 1818 at the age of 22, and promptly set off to study at the famous European centres in Berlin, Göttingen, Vienna, Copenhagen, Paris and Italy. His travels were not without interest and excitement. A facility for foreign languages landed him in an Austrian prison for ten days on the suspicion of being a German spy. While travelling through the Mount Cenis pass in the Alps in the autumn of 1819 he met a young artist and the pair travelled together for some time, neither seeking the other's name. The artist was James Mallard William Turner.

Graves was appointed physician to the Meath Hospital in 1821 at the age of 25. His opening lecture did little to endear him to his seniors. He claimed that many fatalities resulted from indifferent treatment, and he deplored the attitude of medical students who walked the wards in pursuit of entertainment rather than medical knowledge. Graves had been much impressed by the method of bedside clinical teaching on the continent, especially in Germany. He praised the gentleness and humanity of the German physicians who, unlike their Irish and English colleagues, did not have 'one language for the rich, and one for the poor', and whose practice it was to put unpleasant diagnoses into Latin, rather than upset their unfortunate patients.

Graves was joined by his young colleague William Stokes in 1826 and together they began to reform clinical practice in the hospital. Their revolutionary methods caused some resentment, but it is to the credit of the Meath Hospital that it permitted its young physicians to effect their reforms. It was not long before Graves and Stokes had an international reputation that was later acknowledged by the great William Osler, who said: 'I owe my start in the profession to James Bovell, kinsman and devoted pupil of Graves, while my teacher in Montreal, Palmer Howard, lived, moved and had his being in his old masters, Graves and Stokes'.

In 1843, Graves published his famous *Clinical lectures on the practice of medicine,* which was subsequently translated into French, German and Italian. In this book we find evidence of the gift that was common to these Victorian masters of clinical expression — the ability to describe their observations in clear and lively prose. It was the French physician Trousseau who proposed that the illness exophthalmic goitre, described in the *Lectures,* be named 'Graves' disease'.

Graves was revolutionary in his treatment of patients with fever; he advocated supportive therapy rather than starvation, bleeding and blistering. The story goes that one day on his rounds he was struck by the healthy appearance of a patient recently recovered from severe typhus fever and said to his students: 'This is the effect of our good feeding, and gentlemen, lest when I am gone you may be at a loss for an epitaph for me, let me give you one in three words: "He fed fevers"'.

During his professional career he received many honours. He died from cancer of the liver in 1853 at the age of 57.

Born: Thomas Street, Dublin 1802

Died: Merrion Square, 1 February 1880

Family: Son of John Corrigan, merchant, farmer, shop-keeper, chapman and collier-maker, and Celia O'Connor, a descendant of the clan O'Connor

Addresses:
Merrion Square, Dublin
'Inniscorrig', Coliemore Road,
Dalkey

Distinctions:
Graduated M.D. Edinburgh University 1825
Physician to the Charitable Infirmary,
   Jervis Street 1830
Physician to House of Industry Hospitals 1840
Physician-in-ordinary to Queen Victoria in
   Ireland 1847
Senate of the Queen's University 1850
President of King's and Queen's College of
   Physicians 1859–63
Baronet of the Empire 1866
Member of parliament for the City of Dublin
   1870
Vice-chancellor of the Queen's University 1871
Corresponding member of Académie de
   Médicine de Paris.

Dominic Corrigan was the first of the Catholic middle-class to rise to fame in Dublin medicine. Appointed to the staff of the Charitable Infirmary in the year after Catholic emancipation, he quickly fulfilled the promise he had already shown with his famous paper on 'Permanent patency of the aortic valve', a condition which now bears the eponym 'Corrigan's disease'.

As a leading member of the Central Board of Health during the Great Famine, he incurred the wrath of his professional colleagues, most notably Robert Graves, for making what was considered to be a derisory five-shilling-a-day award to doctors working in the famine areas. He was black-balled ignominiously for the honorary fellowship of the College of Physicians but was later made president of the College for an unprecedented term of five years, during which time he instigated the building of the College Hall in Kildare Street. Government rewarded his commitment to the famine cause by making him physician-in-ordinary to Queen Victoria in Ireland and later a baronet of the Empire.

He was a member of the Senate of the Queen's University and a commissioner for national education, and in these roles he advocated much needed reform in education.

He was elected a Liberal M.P. for the City of Dublin in 1870. At Westminster he supported the temperance cause, but his greatest efforts were directed to the issue of university education in Ireland. With remarkable courage and foresight he advocated non-denominational national university education believing that race and creed should not be considerations in third-level education. This stance brought him into bitter conflict with the hierarchy of his church. Corrigan refused to compromise his liberal principles for what he saw to be doctrinaire Catholicism. He feared that the Protestant religious bigotry to which Ireland had been subjected for so long might, with emancipation and the disestablishment of the Church, be replaced by an equally pernicious form of religious intolerance. This conflict with the Catholic Church left him disillusioned and he did not seek re-election to parliament.

He died on 1 February 1880 and was interred in the family vault in St Andrew's Church in Westland Row.

Eoin O'Brien
Royal College of Surgeons in Ireland
Dublin

Kiltorcan Old Quarry.

The small town of Ballyhale lies in southern County Kilkenny. To the east of the town, across the line of the railway that links Kilkenny and Waterford, there rises a low, north to south extending ridge, which occupies the country between the Little Arrigle River and the Arrigle River. To follow the road that runs along the western margin of the ridge was once a pleasurable experience because of the prospects it afforded both of the lowlands around Ballyhale and of the more distant forms of Slievenamon and the Comeragh Mountains. Today, sadly, the immediate environs have in many places become horribly blighted. At various times quarries have been opened up in the country on either side of the road, but today most of the quarries are abandoned and they are being used as refuse dumps. Around the paraphernalia left by former quarrymen are rusting inverted wrecks of abandoned cars, corpses of refrigerators with their guts hanging out, mangled remains of washing machines which will never again wash whiter than white, and masses of trash spilling from plastic bags. The air is filled with a pungent odour which the charitable might attribute to the goats residing in the former quarries, but which the realist will attribute to decaying rubbish in general and, more specifically, to the site which is inscribed 'Septic Tank Disposal'. It is as blighted a piece of rural Ireland as one could ever regret having discovered and for a geologist it is all particularly sad because upon that ridge there lies one of Ireland's most famed fossiliferous localities.

The rocks forming the ridge are sandstones and shales dating from the Upper Devonian Period and laid down about 350 million years ago. Within the sandstones and shales there are preserved the forms of some of the plants and animals which inhabited this part of the world at that distant time. Such fossils have been discovered in the strata at various of the quarries opened up in the ridge but the very small quarry where the fossils were first found has always been held in especial esteem by geologists because of its priority and because it has proved to be peculiarly rich in its fossil contents. Today it is referred to as 'Kiltorcan Old Quarry', Kiltorcan being the townland in which it lies; it is located in a field 1.6 kilometres to the south-east of Ballyhale. The site is located upon the Six Inch (1:10,560) map of County Kilkenny, sheet 32, National Grid reference S557345.

The Kiltorcan fossils were discovered in 1851 by James Flanagan, one of the fossil collectors attached to the Geological Survey of Ireland. At the time the Survey was mapping the rocks of County Kilkenny and it was Flanagan's task to search the strata for fossils because fossils are of the highest significance in the dating of any geological formation. Flanagan was extremely good at his job. By 1851 he had been a professional fossil collector in Ireland for almost twenty years and he had developed a remarkable instinct for finding fossiliferous localities. Flanagan must have been delighted with his first finds at Kiltorcan and thereafter he must have spent many happy hours hammering lumps of rock out of the quarry face and then splitting the rocks open in the hope that some new and handsome fossil would be revealed.

Flanagan reported his discovery to his superiors and the news speedily brought to the site two of the most eminent geologists of the day: Joseph Beete Jukes (1811–69) (Vol. 1, p. 92), the Local Director of the Geological Survey of Ireland, and Edward Forbes (1815–54), the London-based palaeontologist for the entire Geological Survey of Great Britain and Ireland. Both men were fascinated by what Flanagan had to show them; nothing like the Kiltorcan fossils had up to then been found anywhere else in the world. It was Forbes who first brought the fossils to the notice of the scientific world, a duty which he performed in a paper read to the Geological Section of the British Association for the Advancement of Science meeting at Belfast in September 1852.

The sandstones and shales at Kiltorcan are yellowish green and their fossils consist of an assemblage of finely preserved plant remains, some arthropods, a freshwater bivalve, and some fish. Of these it is the plant remains which hold for us the greatest interest, partly because of their intrinsic beauty, partly because they are the remains of very early plants, and partly because we can now see them to possess high relevance to our understanding of the evolutionary story of plants. Most famed among the various types of Kiltorcan plant remains is the fern-like *Archaeopteris hibernica* (Forbes) Dawson 1861: some of the Kiltorcan specimens have fronds about a metre long. But the Kiltorcan beds may still contain exciting and as yet undiscovered secrets. Only in recent years, for example, has it come to be realised that the site probably contains the remains of the earliest seed plants known within these islands.

Forbes's paper of 1852 made the geological world aware of the palaeontological riches of Kiltorcan and since then many a geologist has followed in Flanagan's footsteps to the site in the hope of discovering further palaeontological riches. Repeatedly over the decades money has been spent to bring down substantial quantities of rock from the quarry-face so that geologists might sift through the resultant debris in search of fossils. The most recent event of this type occurred in 1976 when, with the assistance of Roadstone Ltd, parties from Birkbeck College London, University College Bangor and Trinity College Dublin re-explored the Old Quarry under the direction of Professor W. G. Chaloner. Flanagan himself revisited the site several times after his 1851 discovery. In the autumn of 1858, the Geological Survey sent Flanagan back to Kiltorcan to collect more fossils and he found lodgings in the household of Richard Rothe in Ballyhale. Each morning he carried his hammers up the slope to Kiltorcan and each evening he returned laden with the day's finds. But gradually his health began to fail. By March 1859 he was bed-ridden and he died on 14 April 1859. As one of his colleagues quaintly expressed it, Flanagan 'had carried his hammer to another world'. The funeral took place two days later and he evidently lies within sight of Kiltorcan in an unmarked grave in the Roman Catholic burial ground at Ballyhale. After his death his only effects of note were found to be 'a gold watch and chain of very inferior workmanship and material' and a pistol. I wonder why he possessed a pistol.

Further reading:
W. H. Baily, 'On fossils obtained at Kiltorkan Quarry, Co. Kilkenny'. *Report of the British Association for the Advancement of Science, Exeter,* Part 1 (London, 1869), 73–5.
E. Forbes, 'On the fossils of the Yellow Sandstone of the south of Ireland'. *Report of the British Association for the Advancement of Science, Belfast,* Part 2 (London, 1852), 3.
G. L. Herries Davies, *Sheets of many colours: the mapping of Ireland's rocks 1750–1890* (Dublin, 1983).
C. H. Holland (ed.) *A geology of Ireland* (Edinburgh, 1981).

Gordon Herries Davies
Trinity College
Dublin

Each year tens of thousands of visitors to County Kerry must follow the road from Tralee, along the northern shores of the Dingle Peninsula via Stradbally, over the Conair Pass, and down into the town of Dingle. As they travel — Atlantic mists permitting — they will enjoy some of Ireland's finest scenery, but scarcely any of those visitors realise that their route is taking them past one of the classic sites of Irish geology. That site is Lough Doon. The lough is a small, circular lake sitting in a deep and secluded hollow on the north-western flanks of Slievanea just one kilometre to the north-east of the viewing-point in the Conair Pass. The National Grid reference for the lough is Q504006, and an easy climb of just a few hundred metres up the slopes to the east of the Conair Pass road will bring the visitor to the shores of the lough itself. To stand there is a rewarding experience, the more so if the historical significance of the site is understood. It owes that significance to the activities of a young Irishman named John Ball (1818–89).

Ball was born in Dublin, the son of an Irish judge and his County Waterford wife. Even as a child, Ball displayed a deep interest in science and a remarkable facility for making field-observations in natural history. He was educated at St Mary's Roman Catholic College at Oscott in England and then at Christ's College Cambridge and, after leaving Cambridge in 1839, he spent the greater part of the next six years on the continent of Europe travelling and studying natural history. Of particular relevance to the story of Lough Doon is the fact that in 1845 Ball made a prolonged visit to Zermatt in Switzerland and there he investigated the local glaciers in what is a veritable glaciologist's paradise. But that year saw the start of the Irish potato famine and Ball clearly felt a duty-call to make himself useful at home. Back in Dublin, in 1846, he was appointed an assistant Poor Law Commissioner and immediately despatched to County Kerry, there to assist with famine relief measures.

While in Kerry he continued to be a painstaking observer of natural phenomena and one day he must have made an ascent into the basin of Lough Doon. Here a little background information has to be provided. In 1837 the great Swiss naturalist Louis Agassiz (1807–73) had arrived at the startling conclusion that in the recent geological past a large proportion of the northern hemisphere had been submerged under immense glaciers during an episode of earth-history which he proposed to term 'die Eiszeit' — the Ice Age. During a tour of Great Britain and Ireland in 1840 Agassiz had pointed out to local geologists the wealth of field-evidence available to support his belief in the former existence of glaciers within the British Isles, and during the 1840s British geologists gradually began to accept the truth of his interpretation. In Ireland, however, there was some lingering uncertainty on the matter. On his visit to Ireland, Agassiz had himself pointed out features which he claimed to be the legacy of former Irish glaciers in counties Cavan, Down, Dublin and Fermanagh, but could the Emerald Isle really once have been a land held in an icy glacial grip? On 9 March 1842 Charles William Hamilton (c. 1802–80) did report to the Geological Society of Dublin his discovery of relics of former Irish glaciers in the Dingle Peninsula and on Keeper Hill in County Tipperary, but his studies were never published, and when Ball went down to Dingle in 1846 the former existence of Irish glaciers was still a moot point. Ball arrived in Dingle with the landscapes of the country around Zermatt still very fresh in his memory and almost everywhere he looked in County Kerry he saw landforms comparable to those which he knew so well in Switzerland. There was only one difference: in Switzerland the glaciers responsible for shaping the landforms were still present; in County Kerry the glaciers were gone.

Ball perhaps first visited Lough Doon during the autumn of 1846 and he was certainly there again during October 1848. As a result of his observations made during those visits, Lough Doon became the first site in Ireland where detailed investigations proved beyond all doubt the former existence of an Irish glacier. At Lough Doon, Ball noted the amphitheatre and tarn of what would today be termed a glacial corrie or cirque; he noted the glacially rounded, polished and striated rocks forming the rock-bar that impounds Lough Doon; he noted the large numbers of glacially transported boulders; and he noted the fine lobate moraine which runs from the corrie and down the slopes below the Conair Pass road where the moraine forms a dam holding up the small lake of Lough Beirne. Visit Lough Doon today and you will see all the features just as Ball saw them almost a century and a half ago.

Ball reported his observations at a meeting of the Geological Society of Dublin held on 14 November 1849 at the Society's room in House 35 in the New Square of Trinity College. The paper was published in the Society's journal the following year. Ball made no further contribution to the study of Ireland's glacial history but he never lost his interest in glacial phenomena and, when the Alpine Club was founded in 1857, he became the Club's first president. He was the pioneer of that distinguished school of Irish Pleistocene glacial studies which has had among its members figures such as Maxwell Henry Close (1822–1903), William Bourke Wright (1876–1939), John Kaye Charlesworth (1889–1972), Anthony Farrington (1893–1973) (see p. 64), George Francis Mitchell (born 1912), and Francis Millington Synge (1923–83). The accounts of Ireland's glacial history published over the decades by distinguished geologists such as these have all been grounded upon detailed field-studies of precisely the same type as that which Ball conducted at Lough Doon during hours snatched from his official duties as a Poor Law officer during the 1840s.

Further reading:
J. Ball, 'Notice of the former existence of small glaciers in the County of Kerry'. *Journal of the Geological Society of Dublin* **4** (2) (1849), 151–4.
For biographical information on Ball see the *Dictionary of national biography* and *Geological Magazine*, new series, **7** (1) (1890), 47–8.

Gordon Herries Davies
Trinity College
Dublin

At Rathfarnham, five miles south of the centre of Dublin, Archbishop Loftus built a castle in 1585. Having been extensively 'Georgianised' and passing through several hands, it was purchased by the Society of Jesus in 1913 and extended to form a Jesuit house of studies. It is now the property of the nation. For about half a century, the castle was the centre of seismology, the study of earthquakes, in Ireland.

Following the San Francisco earthquake of 1906, several Irish Jesuits, in particular Father H.V. Gill, became interested in the distribution of earthquakes around the world, and in 1909 three Jesuit Colleges, Stonyhurst in England, Riverview in Australia and Mungret near Limerick, set up seismological observatories, using instruments by Mainka of Strasbourg (though no record or remains are known to exist of those at Mungret). The Mungret observatory was in charge of Father William O'Leary, S.J. He was born in Dublin in 1869, the son of Dr W.H. O'Leary, professor of anatomy at the Royal College of Surgeons, and entered the Jesuit Order in 1886. At Mungret, in addition to looking after the seismographs and teaching, he carried out for the Royal Meteorological Society a study of the upper atmosphere using sounding balloons, this being the most westerly station in Europe.

In 1915 Father O'Leary was transferred to Rathfarnham. He already had constructed, while at Mungret, a seismograph of his own design, and this was exhibited at the Coronation Exhibition in London in 1911. At the time of his transfer he was working on a much larger one, which became operational at Rathfarnham in 1917.

To record an earth movement it is necessary to have a mass freely suspended so that its inertia will keep it at rest but, in order to obtain a sensitive record, the suspended mass should be able to swing freely with a natural period much longer than the time scale of the initial tremor. O'Leary achieved this by using an inverted pendulum. The 'bob' consisted of about 1800kg of octagonal iron plates, the top being just above ground level. The rod, a little over two metres long, ran down into a concrete pit below it. At the bottom of the rod was a steel disc, to the rim of which were attached three steel cables making an angle to the vertical and secured to the ground at the top of the pit. This pendulum, nearly unstable and thus very sensitive, had a period of almost sixteen seconds. Electromagnetic dampers brought the pendulum quickly to rest to record the next tremor. Two systems of levers magnified the movements in two dimensions and recorded them through glass styli on a smoked paper drum driven by clockwork. Time marks were recorded electro-magnetically.

It is sometimes said that O'Leary built the inverted pendulum seismograph entirely with his own hands. In view of the size of the apparatus this must be an exaggeration, and it is almost certain that most of the instrument was actually made by the firm of Grubb (see p. 20), the foremost instrument makers in Dublin at that time. This is in no way to belittle the achievement of the conception, theoretical analysis, design and commissioning of a pioneering, large and successful piece of apparatus by a priest who had many other duties. Grubb certainly built parts for a photographically recording three-component (two horizontal, one vertical) seismograph designed by O'Leary in 1913. Construction was delayed by the Grubbs' involvement in

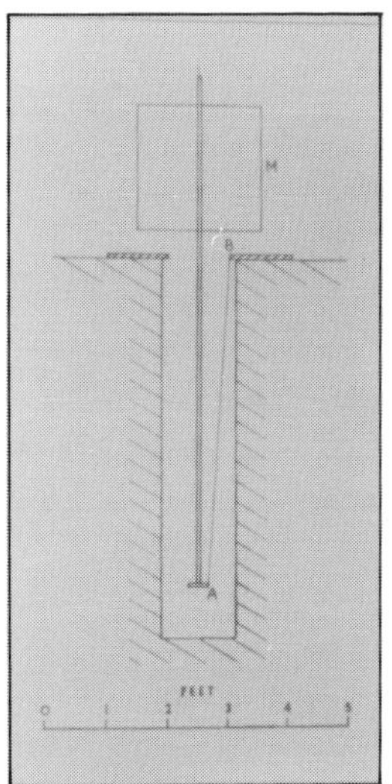

Inverted pendulum.

The 'big O'Leary'.

war work and transfer from Dublin to St Albans and, although a foundation pillar was installed at Rathfarnham and the instrument assembled, there is no record that it was completed or used. Possibly by then it was too late, since in 1929 Father O'Leary left Ireland to direct the Riverview College Observatory, New South Wales, where there is still a mercury compensated clock designed, patented and constructed by him. Here he continued seismographic work, but also turned his attention to the photographic study of variable stars. Once again he designed and built apparatus, in this case a blink comparator. He died there in 1939.

The work at Rathfarnham was continued by Father Richard Ingram, S.J., who also lectured in mathematics at University College Dublin, and by Father P. M. Troddyn, S.J. The O'Leary seismograph continued in use, but a Milne-Shaw seismograph was added in 1932. This used a horizontal pendulum, resembling the boom of a yacht, about 0.4m long, of mass about 0.5kg. Its outer end was supported by two fine steel wires from the top of a pillar, near the base of which a pin on the inner end of the boom rested in an agate cup. The instrument was electromagnetically damped and movements were recorded by an optical beam and mirrors giving a magnification of 250, on photographic paper.

Routine recording continued on both instruments until Father Ingram died in 1967, after which the observatory fell into disuse. The Milne-Shaw instrument was removed for use elsewhere, but the 'big O'Leary' was allowed to decay, a sad end to a noble achievement in Irish science. It was finally dismantled in 1979 and most of the parts are in the museum at St Patrick's College, Maynooth. Most of the three-component machine is lost. Seismology is now in the hands of the Dublin Institute for Advanced Studies at Merrion Square, whose former director, Professor Thomas Murphy, I thank for much assistance.

Further reading:
R.E. Ingram and J.R. Timoney, 'Theory of an inverted pendulum with trifilar suspension'. *Geophysical Bulletin* No.9 (Dublin Institute for Advanced Studies, 1954).
R.E. Ingram and P.M. Troddyn, 'Earthquake recording at Rathfarnham Castle'. *Irish Jesuit Directory and Year Book* (1938), 139.
Rev. T. MacMahon, 'Rathfarnham Castle'. *Dublin Historical Record* **41** (1987), 21–3.

Adrian Somerfield
St Columba's College
Whitechurch, Dublin

Leinster House, headquarters of the Society from 1815 to 1924 (from a Malton print, 1792).

The Royal Dublin Society (RDS) was not set up as a primarily scientific society. At its first meeting on 25 June 1731, the objectives of the society were set out as the improvement of 'Husbandry, Manufactures and other Useful Arts'. On reflection, the words 'and sciences' were added to these at the next meeting.

It was early days for science in its modern sense. The emphasis of the Society was on usefulness to the prosperity of the country, which to a certain extent would correspond with the personal interests of the founders. They were sufficiently abreast of the times to realise, after some thought, that science could be useful in this way. This was a reflection of the growing impact of the thinking of people like Robert Boyle (1627–91) from Lismore, County Waterford, who had challenged the traditional emphasis of science as a solely intellectual discipline. His contemporaries, like Christian Huygens (1629–95) and Gottfried Leibniz (1646–1716), were inclined to doubt the value of demonstrating by experiment what they and all 'rational thinkers' knew to be true by logical reasoning.

In fact, the promotion of science has turned out to be a most important part of the work of the Society over its history, although its place in the Society's programme has waxed and waned. Tension between the demands of agriculture and science, with sometimes one and sometimes the other gaining precedence, has been a feature of the Society's activities. At its best, science and agriculture have been enhanced together since, of course, science has had an enormous part to play in the development of agriculture.

Six aspects of the Society's activities can be picked out to illustrate its impact over the years: the beginning of useful work that would later be developed through national institutions; the giving of premiums and awards; the organisation of lectures and instruction; the opportunities and encouragement given to its employees; the publication of journals and books; and the promotion of discussion and debate on science policy and practice.

National institutions:

Often, to get something useful going, solid groundwork is needed. Governments can be so busy managing the day-to-day affairs of state that there is no time to sit and reflect on what would be good for the country in the long term. The Society, by purchase and by receiving gifts of books, built up an important collection, which became the National Library of Ireland under the terms of the Dublin Science and Art Museum Act of 1877. From its earliest days, it carried out experiments in plant cultivation and so was the obvious choice when the idea of setting up a botanic gardens was mooted around 1790. The gardens were taken over by the state, also in 1877, as was the Society's zoological collection, now part of the National Museum. It took initiatives which led to the setting up of the Veterinary College, now part of University College Dublin.

In this century, the Irish Radium Institute (Vol. 1, p.102) was founded by the RDS to supply radon to hospitals for the treatment of cancer.

Combining science with culture, its practical drawing schools led eventually to the foundation of the National Gallery and, in recent years, it played a major part in the setting up of the Crafts Council of Ireland.

Premiums and awards:

In its earliest days, the Society offered premiums to encourage activities of benefit to the country. It was Dr Samuel Madden (1686–1765) who started this off in 1739. The list of premiums offered by the Society in 1766 takes up 34 printed pages in the Society's *History* (written by H.F. Berry and published in 1915). Bog reclamation, the growing of cereals and vegetables, tanning, bee husbandry, fisheries, tree planting, horse breeding, iron and steel making, textiles, printing and mining were but some of the activities to win the Society's premiums.

Awards of various kinds have been offered by the Society right up to the present, the best known probably being the agricultural prizes in the Spring Show, and the equestrian prizes in the Horse Show. Since 1899, the Society has awarded its Boyle Medal for 'scientific research of exceptional merit carried out in Ireland'—and this has recognised the achievements of some of the greatest of Ireland's scientists, many of them appearing in the pages of this book and its sister volume.

Lectures and instruction:

While the long established Trinity College Dublin (founded in 1592), and other universities and colleges of more recent date, have of course had among their primary functions the giving of lectures and instruction, these have been generally and necessarily restricted to their own students. For many years high quality instruction for the non-student public was not available.

It was in this area that the Society played one of its major roles in the advancement of science. From the end of the eighteenth century it appointed professors in a variety of disciplines — including botany, chemistry, geology, mineralogy and natural philosophy. These professors were required to give courses of lectures in Dublin open to the public and, in addition, they gave lectures in provincial centres around the country. After 1845 they were transferred to the 'Museum of Irish Industry and Government School of Science applied to Mining and the Arts' in St Stephen's Green, which was eventually subsumed into the Faculty of Science in University College Dublin.

The Society's Boyle Medal.

Encouragement of individuals:
An aspect of the Society's history that has not been given sufficient emphasis is the encouragement it has given to its members or employees in the development of their careers — which, in turn, have greatly benefited the country. Before the time of the professional scientist it gave opportunities to members such as Richard Kirwan (1733–1812: Vol. 1, p. 14), on his way to becoming one of the world's premier chemists and mineralogists. One of the first career chemists was William Higgins (1763–1825: Vol. 1, p. 16) for whom a laboratory was set up by the Society in 1792 to encourage his 'experiments in dyeing materials and other articles, wherein chymistry may assist the arts'.

Sir Richard Griffith (1784–1878: Vol. 1, p. 28) started his scientific career as an employee of the Society and eventually produced his remarkable geological map of the whole of Ireland in 1839. In 1836, Edmund Davy (1785–1857: Vol. 1, p. 105), the Society's professor of chemistry, prepared and described acetylene (which he called bicarburet of hydrogen), 23 years before it was rediscovered by Bertholet, to whom its discovery is commonly attributed. Sir Robert Kane (1809–90: see p. 24 of this volume) was professor of natural philosophy to the Society from 1834 to 1847, during which time the Society published his *Natural resources of Ireland*. He went on to become a fellow of the Royal Society and president of Queen's College Cork. It was in the Society's laboratory that James Emerson Reynolds (1844–1920: Vol. 1, p. 58) prepared thiourea (the sulphur equivalent of urea) for the first time — a task which had defeated many other chemists. Reynolds went on to become professor of chemistry in Trinity College Dublin in 1875. More recently, Horace H. Poole obtained the Boyle Medal of the Society in 1936, much of his work on photoelectricity being carried out while he was employed by the Society, first as chief executive officer for science and later as registrar.

Journals and books:
The great advantage of the printed word over other less tangible aspects of the promotion of science is that it survives in what is now called 'hard copy'. There can be little doubt of the influence the RDS has had on science in Ireland, but nowhere can this be better attested than in the quantity and quality of its publications.

Between 1800 and 1985, it published nine journal titles, containing some 2436 entries in all, many of them of special importance in the development of scientific thought. To take just one example, it was in a paper by George Johnstone Stoney (Vol. 1, p. 50) in *The Scientific Transactions of the Royal Dublin Society* in 1891 that the term 'electron' was first used. The Society's many published books have included statistical surveys of the counties of Ireland, 22 of these being published between 1801 and 1829.

It is perhaps appropriate that, recently, the Society has instituted a series of books with the general title 'Historical Studies in Irish Science and Technology' in which eight volumes were published between 1980 and 1989. In fact, in cooperation with the Royal Irish Academy, the Society is helping to develop a greater appreciation of the history of Irish science — previously a much-neglected topic. It is this cooperation which has led to the publication of these books on people and places in Irish science and technology. In 1985, the two organisations set up a Joint Committee on Historic Scientific Instruments, of which Ireland has been found to have surprising riches, particularly of the nineteenth century, and three RDS publications have already resulted.

Discussion and debate:
Since it ceased in the late nineteenth century to receive its major funding from government sources, the Society has been proud of its independence. It claims with justification to be an 'honest broker', since it exists to promote the welfare of the country rather than that of its members. This can be a valuable asset in a country which has seen the great influence of pressure groups and special pleading over the years.

One way in which the Society has used this position is in the organisation of seminars and debates in various scientific areas, usually publishing the results. Recent examples include: 'Genetics and horse breeding' and 'The Bull Island' (1975); 'A profit and loss account of science in Ireland' (1981); 'Air pollution in Ireland: Dublin — a case study' (1985); and 'The utilisation of Irish midland peatlands' (1988).

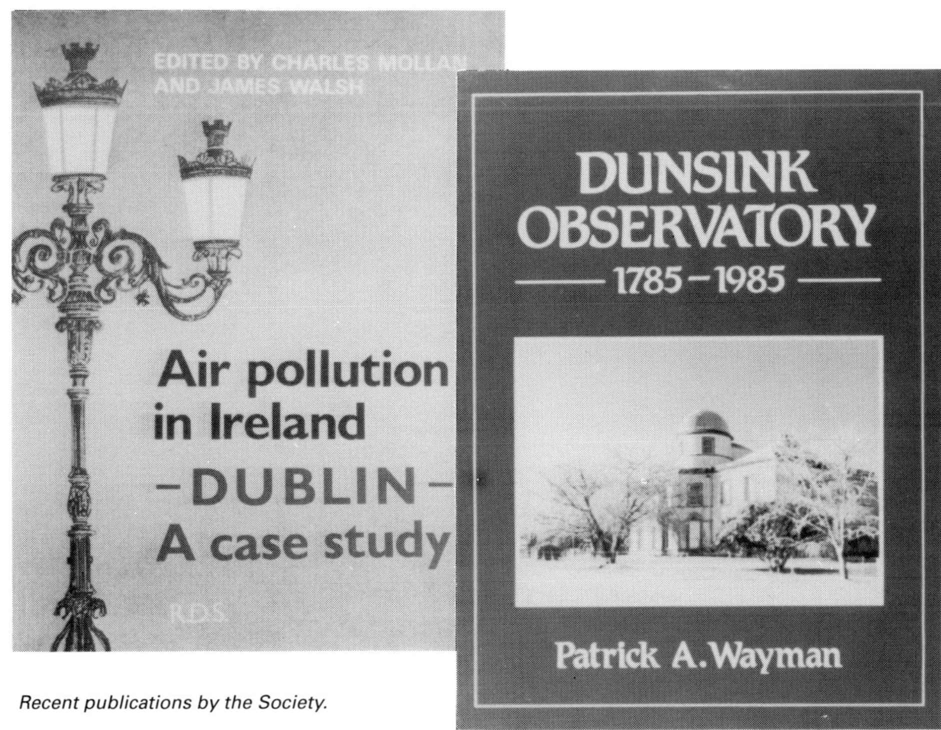

Recent publications by the Society.

Further reading:

H. F. Berry, *A history of the Royal Dublin Society* (London, 1915).
T. de Vere White, *The story of the Royal Dublin Society* (Tralee, 1955).
J. Meenan and D. Clarke (eds), *RDS — The Royal Dublin Society 1731–1981* (Dublin, 1981).
C. Mollan, *Nostri Plena Laboris — an author index to the RDS scientific journals, 1800–1985* (Dublin, 1987).

Charles Mollan
Royal Dublin Society
Ballsbridge
Dublin

Dawson Street, Dublin; Academy House on the right.

The Mansion House, located in Dublin's Dawson Street, must be familiar to every Dubliner. In contrast, there are probably very few people who know anything of the activities which centre upon the handsome, brick-fronted building immediately to the north of the Mansion House. That building — number 19 Dawson Street — dates from about 1755 and since 1851 it has been the home of the Royal Irish Academy, Ireland's premier learned society. The Academy was founded in 1785 during a period when similar scholarly bodies were being established in a number of cities and towns within these islands. The Academy's founders, led by the earl of Charlemont (1728–99), felt that it was in the national interest to establish an institution where gentlemen might convene in the pursuit of all branches of scholarly learning, and the wide scope of the Academy's interests is reflected in the title which its founders bestowed upon their creation: the Royal Irish Academy for Science, Polite Literature, and Antiquities. The motto chosen for the Academy was simple: 'We will endeavour'.

Originally the Academy was housed at 114 Grafton Street (the building no longer exists) but, whether in Grafton Street or Dawson Street, the Academy has always devoted itself to the furtherance of all branches of learning from archaeology and astronomy to philology, philosophy, topology and zoology. In view of its concern with the entire field of scholarship, there are many Academy activities which must pass unnoticed in a volume focused only upon Irish science and technology. What follows is therefore a survey of but a part of the Academy's work and the reader must remember that, in addition to its many involvements with science, the Academy is also a major centre for research in the humanities. In particular it must be mentioned that the Academy possesses a library of international significance containing material of importance to scholars investigating almost every aspect of Irish culture.

If, with your interest in science, you were to enter Academy House expecting to find white-coated scientists watching dancing patterns upon oscilloscopes, or feeding abstruse mathematical hieroglyphics into computer terminals, then you will be disappointed. The Academy is not a centre for research in science at the level of the laboratory bench. The Academy makes its contribution to Irish science — to international science — at a higher and more general level and here there must be mentioned first the Academy's role in publication. All scientists are communists in the sense that, within the scientific community, there is no such thing as private scientific knowledge. All scientific knowledge has to be disseminated and shared universally through the medium of publication. The Academy assists in this international sharing process by publishing journals through which Irish and other scientists may bring their research to the international community of science. The Academy's earliest journal was the *Transactions of the Royal Irish Academy,* which appeared in 33 volumes between 1787 and 1907, but today the Academy's chief publication is the *Proceedings of the Royal Irish Academy* founded in 1836. Almost 90 volumes of the *Proceedings* have now been published and that portion of the journal devoted to science consists of two parts: Section A, mathematical and physical sciences; and Section B, biological, chemical and geological science. In addition, since 1984 the Academy has also published the *Irish Journal of Earth Sciences,* a journal originally founded by the Royal Dublin Society in 1978. Through these journals the Academy has published thousands of scientific research papers, many of them being papers possessed of considerable significance. There was Richard Kirwan (Vol. 1, p. 14) on 'the supposed igneous origin of stony substances' (*c.* 1795), John Brinkley (this Vol., p. 102) on 'the measured parallax of certain stars' (1815), Sir William Rowan Hamilton (Vol. 1, p. 36) on 'a new species of imaginary quantities, connected with a theory of quaternions' (1843), Robert Mallet (Vol. 1, p. 42) on 'the dynamics of earthquakes' (1848), and the 67 natural history papers published in the *Proceedings* volume 31 (1911–15) and which present the results of the famed Clare Island Survey (Vol. 1, p. 94). The Academy has exchange agreements with many academies overseas and, in return for the *Proceedings*, it receives the publications of the foreign academies. In this manner, and at trifling cost, the Academy has built up a fine collection of foreign scientific periodicals.

A second manner in which the Academy stimulates science is through the organisation of meetings at which scientists can meet to discuss their achievements and problems. Some of these meetings take the form of a discourse delivered by some distinguished visiting scientist. Other of the meetings are symposia extending over several days, during which a number of scientists will read papers describing their most recent research. Through the staging of such events, the Academy provides a meeting place for scientists coming from many different backgrounds and institutions, and it thus affords valuable opportunity for the cross-fertilisation of scientific ideas. But the scholars sometimes like to gather in a yet more convivial setting, and the Academy therefore possesses a Dining Club which traces its origins back as far as 1826. At meetings of the Club, as the roast beef is carved and as the port circulates, perhaps not *all* of the

The Academy's Cunningham Medal, first awarded in 1796 to Thomas Wallace and, most recently, in 1989 to G. F. Mitchell.

Academy House, the Meeting Room.

conversation relates to phenocrysts, phyllopods and pulsars, but the Club does provide members of the Academy with yet another opportunity of becoming better acquainted with each other and with the varied problems that are engaging attention at the frontiers of science.

Academy House may not itself be a centre for actual scientific investigation but, ever since its foundation, the Academy has from time to time made monies available for the support of scientific research programmes. Sadly, such monies have never been available upon anything more than a very modest scale, but this kind of expenditure has allowed the Academy to play a very direct role in the shaping of Irish research endeavour. As early as 1788, the Academy sought to establish a national network of meteorological observatories, and it voted £30 for sets of barometers and thermometers, although in the event little came of the project. Far more successful was the Academy's second involvement with Irish meteorology when, in 1851, sixteen stations spread throughout the country were equipped with a variety of instruments at a cost to the Academy of £225. Later research projects which have received Academy funding include the excavation of fossils from Ballybetagh Bog (Vol. 1, p. 78) in the Dublin Mountains, an Academy dredging expedition to Rockall in 1896, a joint Royal Dublin Society/Academy expedition to Plasencia in Spain to observe the 1900 eclipse of the sun, and the Clare Island Survey which was launched at a meeting held in the National Museum on 23 April 1908 attended by thirteen gentlemen, nine of them members of the Academy. More recently, and to mark its bicentenary, the Academy has established a Bicentennial Fellowship to which scholars of proven research ability are appointed for a period of two or three years to allow them to bring some project to the stage of publication. From time to time that Fellow will presumably be a scientist. While dealing with this issue of funding, one question is bound to arise: from what source does the Academy obtain its income? The answer is simple. The Academy's principal source of income is an annual grant-in-aid provided by the Higher Education Authority.

The Academy represents Ireland in all negotiations with similar scholarly institutions overseas, and the Academy's relationships with the Royal Society of London (founded 1660) and the Royal Society of Edinburgh (founded 1783) are particularly close. Through these relationships, the Academy is able to arrange the international exchange of scientific personnel, the organisation of international scientific

Extract from an Irish astronomical manuscript (Stowe, B.II.1.§19) perhaps dating from about 1400 and now in the library of the Royal Irish Academy. The text begins: Si autem sol minoris esset cainnditatis . . . (If the size of the sun were less than that of the earth . . .). (The text is largely a translation of a work in Arabic by Messahalah (Fl. c. A.D. 800).

symposia, and Irish participation in international research programmes. Another way in which the Academy forges links with the international scientific community is through the election of distinguished foreign scientists to honorary membership of the Academy. In this manner many of the greatest of the world's scientists have been brought into association with the Academy, among their number being Louis Agassiz (1863), Niels Bohr (1933), Albert Einstein (1928), Enrico Fermi (1944), Werner Heisenberg (1964), Alexander von Humboldt (1849), Justus von Liebig (1837), Dmitry Mendeléeff (1889), Ivan Pavlov (1917) and Max Planck (1911). Since 1955, the Academy has enormously expanded the scope of its activities, both at home and overseas, through the establishment of national committees, each national committee being concerned with one of the specialist areas of scholarly endeavour. At present in science the Academy administers twelve such national committees, their areas of responsibility being as follows: astronomy, biochemistry, biology, chemistry, engineering sciences, geography, geology, history and philosophy of science, mathematics, nutritional sciences, physics, and theoretical and applied mathematics. The membership of the national committees is drawn from among the members of the Academy itself and from designated bodies in all parts of Ireland.

There are today about 250 members of the Academy, and among their number are to be found most of Ireland's leading scholars. A woman scientist — Mary Somerville — was elected an honorary member of the Academy as far back as 1834, but it was not until 1949 that the first woman scientist — Phyllis Clinch — was elected to ordinary membership of the Academy. Each year, the members of the Academy meet to elect eight new ordinary members, the statutes requiring that four of the eight shall be scientists. The election takes place upon the eve of St Patrick's Day (unless the eve happens to fall upon a Sunday), and there are always many more candidates than the eight places available. Competition for election is keen: such is the national and international standing of the Academy that any scholar is proud to be able to add the initials M.R.I.A. after his or her name.

Further reading:
The Royal Irish Academy: a brief description (Dublin, 1978).
T. Ó Raifeartaigh (ed.), *The Royal Irish Academy: a bicentennial history 1785–1985* (Dublin, 1985).

Gordon Herries Davies
Trinity College
Dublin

The 1785 main building of Dunsink Observatory, architect Graham Moyers.

Dunsink Observatory, five miles north-west of the centre of Dublin, stands at a height of 85m above sea level on a low hill that commands a fine view over Dublin and the country to the north. The site was chosen by Rev. Henry Ussher, who was appointed in 1783 to be the first Andrews' professor of astronomy in the University of Dublin. The professorship was founded by the will of the provost of Trinity College, Francis Andrews (provost 1758–71), who left money for building an astronomical observatory for the college and for endowing the professorship. Much later, in the 1930s, the enthusiasm of Eamon de Valera for the tradition of William Rowan Hamilton (Vol. 1, p. 36) at Dunsink was sufficient to ensure that the observatory became part of the Dublin Institute for Advanced Studies when the School of Cosmic Physics was founded in 1947.

The work of the observatory in astronomical observation and research from 1785 to 1921 is that of successive Andrews' professors, who, from 1792, received the courtesy title Royal Astronomer of Ireland. Their dates in office were:

Rev. Henry Ussher	1783–1790	Arthur A. Rambaut	1892–1897
Rev. John Brinkley	1790–1827	Charles Jasper Joly	1897–1906
Sir William Rowan Hamilton	1827–1865	Sir Edmund Taylor Whittaker	1906–1912
Franz Friedrich E. Brunnow	1865–1874	Henry Crozier Plummer	1912–1921
Sir Robert Stawell Ball	1874–1892		

Until 1900, the observational work of the observatory was almost entirely concerned with the positions of stars. The first instruments, in use up to 1865, were made by Jesse Ramsden of London and included an 8-foot graduated circle for measuring the altitude of stars as they crossed the meridian. Transit times were, up to 1875, recorded with two 1787 pendulum clocks by John Arnold, which survive in perfect working order.

From 1870 to 1900, the two principal instruments were a Pistor and Martins transit circle (in use until 1937) and the 12-inch refractor made by Thomas Grubb (this Vol., p. 18) of Dublin in 1852, equipped with a lens made by Cauchoix of Paris in 1829, when it was the largest lens in existence. It was purchased by Sir

James South of London and presented to Trinity College Dublin in 1863 in commemoration of the chancellorship of the third earl of Rosse (Vol. 1, p. 84). This telescope is still in use on public open nights in the winter months and was renovated during the Dublin millennium year of 1988.

The transit circle was used for some extensive observing programmes, especially in the years 1890–1905, when the principal observer was Charles Martin, who was acting director from 1921 to 1936. From 1874 to 1937 this instrument was used to regulate a sidereal clock, which in turn was used to correct a mean-time clock whose beats controlled 'slave' clocks in Dublin. Electrical 'currents' were sent out on telegraph wires to several localities, including principally the Ballast Office of the Port and Docks Board in Westmoreland Street. The time signals were used by ships in the port, about to set out on ocean voyages, to correct and regulate their chronometers: the public dial which was in use for many years in that system, now controlled automatically by radio, is to be seen on the river frontage of the building which has replaced the old Ballast Office building.

Astronomical work always includes theoretical studies based on the principles of mathematics and physics which have been developed over many decades and which still advance rapidly. Astronomy has stimulated and contributed to these developments in many important ways. Contributions of W. R. Hamilton (Vol. 1, p. 36) and of E.T. Whittaker (this Vol. p. 103) at Dunsink were outstanding examples of work done in mathematics at Dunsink in the past.

C. J. Joly, second cousin to John Joly (Vol. 1, p. 64), was also a fine mathematician but he and his successors (particularly H. C. Plummer) carried out work related to the physics of the stars, which began to be understood in the decades 1890 to 1910. They used a 15-inch reflecting telescope erected in the dome of the main building in 1895, part of which was the gift of Isaac Roberts of Liverpool.

This work ceased after Plummer left in 1921, and the Andrews' professorship remained vacant for many years. However, Eamon de Valera, who in 1908 was a professor of mathematics at the Carysfort Teachers Training College in Blackrock, studied mathematics with great enthusiasm at the time when E.T. Whittaker was Andrews' professor. In 1947, when he was taoiseach, de Valera was assisted by Whittaker in initiating legislation to incorporate Dunsink into the School of Cosmic Physics as part of the Dublin Institute for Advanced Studies. The two other sections of the School, at 5 Merrion Square, Dublin, have been active in cosmic ray studies and in geophysics since 1947. The three sections are responsible to the Governing Board of the School and one of the three senior professors serves as director of the School.

Under Professor Hermann Brück (1947–57) and two succeeding astronomy section senior professors, M.A. Ellison and P.A. Wayman, Dunsink Observatory has provided access to modern methods of observation in astronomy, in South Africa at the Boyden Observatory from 1952 to 1976, at the Cape Observatory from 1960 to 1965, and on the Spanish island of La Palma in the Canaries since 1984. Dunsink operates an Irish share in the telescopes of the United Kingdom and the Netherlands on behalf of all interested scientists in the Republic and, in 1988 and 1989, this has included use of the 4.2m William Herschel Telescope, one of the most powerful instruments in the world and one which has the advantage of being installed on a very fine site. Work has also been possible with the space vehicles of the European Space Agency, through Ireland's national membership, and electronic work at Dunsink, begun in the 1950s under H. A. Brück, has contributed to Irish-German space experiments on Giotto in 1986 and to the Soviet Phobos mission to Mars in 1988 and 1989. The scientific work of these 'space probes' has been shared by the Max Planck Institute for Aeronomy (Lindau, W. Germany), the physics department of St Patrick's College, Maynooth, and the cosmic ray section of the School of Cosmic Physics. The geophysics section of the School has pioneered the exploration of Ireland's crustal and oceanic environment, including access to mineral resources, by using geophysical techniques.

Further reading:
'Eamon de Valera centenary' (Dublin Institute for Advanced Studies,1982).
F.E. Dixon, 'Dunsink Observatory and its astronomers'. *Dublin Historical Record* **11** (1950), 33–50.
N.P. O'Hora, 'The Dunsink Observatory'. *The Observatory* **81** (1961), 189–95.
P.A. Wayman, *Dunsink Observatory, 1785–1985* (Royal Dublin Society, 1987).

Rev. John Brinkley, F.R.S., enrobed as Bishop of Cloyne , 1835.

JOHN BRINKLEY   1766–1835
Astronomer

Born: Woodbridge, Suffolk, December (?) 1766

Died: Dublin, September 1835

Family:
Married: Esther Weld, daughter of Matthew Weld of Dublin, 1792
Children: John (b.1793) and Matthew (b.1797)

Distinctions:
Senior wrangler, mathematical tripos,
   Cambridge 1788
Fellow of Gonville and Caius College,
   Cambridge 1789–90
Andrews' professor of astronomy 1790–1827
Ordained priest in the Church of England,
   Lincoln 1791
Created first royal astronomer of Ireland 1792
Fellow of the Royal Society, London 1803
Doctor of divinity, University of Dublin 1806
Cunningham Medal, Royal Irish Academy 1818
Copley Medal, Royal Society 1822
President, Royal Irish Academy 1822-35
Bishop of Cloyne, Co. Cork 1826-35
President, Royal Astronomical Society 1831-34

Addresses:
1790–1827 Dunsink Observatory, Dublin
1827–1835 Bishop's Palace, Cloyne, Co. Cork

Brinkley was chosen by Provost Hely Hutchinson to succeed Henry Ussher, the first Andrews' professor of astronomy in Dublin University, in 1790, and on the advice of Nevil Maskelyne, astronomer royal of England. Although he maintained contact with the Royal Society of London and the Royal Astronomical Society, and has been the only Irish resident to serve as president of the Royal Astronomical Society, Brinkley did all his astronomical work at Dunsink Observatory.

One of the principal problems that Brinkley attacked was that of refraction in the atmosphere, which had been one of Henry Ussher's intended investigations. He measured the constants of aberration and nutation due to the motion of the earth, and he made it his life's work to try to determine the trigonometrical parallax of bright stars. This last part of his work was not successful, owing to instrumental problems, and it involved him in a celebrated controversy with John Pond, Maskelyne's successor at Greenwich Royal Observatory in London. Nevertheless, Brinkley was reckoned to have devised powerful procedures for careful treatment of observational material.

Brinkley made contributions to the mathematics of astronomy and of statistical methods, being one of the first to use least-squares methods to interpret observational readings in practical astronomy, and he did much to introduce 'continental' notation into calculus in Ireland and England.

At the same time as working as an astronomer, Brinkley became an authority in ecclesiastical law in the Church of Ireland, and was eventually created bishop of Cloyne in 1826.

Further reading:
R.S. Ball, *Great astronomers* (Isbister,1895), 233–46.

EDMUND TAYLOR WHITTAKER   1873-1956
Astronomer and Mathematician

Sir Edmund Whittaker, F.R.S.

Born: Birkdale, Lancashire, 24 October 1873

Died: Edinburgh, 24 March 1956

Family: Son of a railway engineer and contractor
Married: Mary Boyd, daughter of the Rev. Thomas
Boyd of Cambridge in 1901.
Children: Three sons and two daughters.

Distinctions:
Second wrangler, mathematical tripos, Cambridge
1895
Fellow of Trinity College, Cambridge 1896
Fellow of the Royal Society 1905; Sylvester Medal
1931; Copley Medal 1954
Andrews' professor of astronomy, University of
Dublin, and royal astronomer of Ireland 1906–12
Professor of mathematics, University of Edinburgh
1912–46
President, Royal Society of Edinburgh 1939–44
The Cross pro Ecclesia et Pontifice 1935
Member, Pontifical Academy of Sciences 1936
Knight Bachelor 1945

Addresses:
1906–1912   Dunsink Observatory, Dublin
1954   48 George Square, Edinburgh

Edmund Whittaker did much of his work in Cambridge, before coming to Dublin in 1906, and in Edinburgh, after leaving Dublin in 1912. Nevertheless, his years at Dunsink Observatory, which he remembered vividly for all his days, were influential in focusing his attention on the problems of incorporating sound mathematics into his astronomy, both as a tool for interpreting observations and as a mode of expression of physical theories. Although he was himself a mathematician, he saw clearly that progress in astronomy required an approach based on theoretical physics and, during his time as royal astronomer of Ireland, he wrote his influential *History of the theories of the aether and electricity* which, in its 1951 second edition, is a standard work on the subject.

At Dunsink, Whittaker undertook responsibility for the practical work of the Observatory, mainly concerned with maintenance of accurate time; and an opportunity was taken to develop satisfactory photometry of stars using the photographic capability of the 15-inch Isaac Roberts reflector. Accurate light curves for bright variable stars were derived, and this work was continued later by Plummer and Martin. It became influential in assisting Eddington, in Cambridge, to develop theories of stellar structure.

Whittaker's period as royal astronomer at Dunsink and his influence on de Valera then and later were major factors in determining the course of scientific work in the Dublin Institute for Advanced Studies up to the present day.

Further reading:
D. Martin and others, 'Whittaker Memorial Number'. *Proceedings of the Edinburgh Mathematical Society*
11 (June, 1958).
W. H. McCrea, 'Edmund Taylor Whittaker'. *Journal of the London Mathematical Society* **32** (1957), 234–56.

Patrick A. Wayman
Dunsink Observatory
Dublin

65 Merrion Square, site of the School of Theoretical Physics, 1940-71

The Dublin Institute for Advanced Studies was created by an Act of the Irish Parliament signed by President Douglas Hyde on 19 June 1940. The Act enabled the government to set up constituent schools in the Institute by Establishment Orders, and such orders were immediately made to establish the School of Celtic Studies and the School of Theoretical Physics. The choice of these two schools reflected the personal interests in mathematics and the Irish language of Eamon de Valera, who was then taoiseach.

Erwin Schrödinger was appointed senior professor in the School of Theoretical Physics and he moved into his office at 65 Merrion Square in February 1941. In June of the same year he was joined by Walter Heitler, who had been appointed assistant professor. Schrödinger and Heitler set about providing, for the benefit of university staff members and senior students, introductory courses on quantum mechanics, which at the time was little known in Ireland. They also conducted seminars on their own current researches. Initially Schrödinger's seminars were devoted largely to his attempts to construct within the framework of general relativity a theory which united gravitational and electromagnetic phenomena. Heitler spoke about his radiation damping theory and its applications to the study of cosmic rays. These seminars of Heitler were responsible for the introduction of cosmic ray research to University College Dublin and for the inclusion of a cosmic ray section in the School of Cosmic physics when it was established in 1947.

During the early years of the Institute, communications with Great Britain and North America were difficult and communications with continental Europe were well nigh impossible. Schrödinger made a major attempt to combat the isolation of the School by holding a colloquium which lasted from 16 to 29 July 1942. The speakers from abroad were P.A.M. Dirac, who delivered five lectures on quantum electrodynamics, and A.S. Eddington, who gave the same number of lectures on unification of relativity theory and quantum theory. These two sets of lectures were published as numbers 1 and 2 of *Communications of the Dublin Institute for Advanced Studies*, Series A.

In the following year a colloquium on crystals was held, the chief speakers being Max Born, P.P. Ewald and Kathleen Lonsdale, and in 1945 a colloquium on topics ranging from quantum electrodynamics to the theory of solids was held with Dirac, Born and L. Jánossy as the main speakers. Thus already during World War Two Irish scientists were able to establish personal contact with some of the leading figures of twentieth-century physics. Then in March 1946 W. Pauli visited the School for two weeks and normal contacts were gradually resumed, but the summer colloquia were still frequently held.

In the establishment order of the School of Theoretical Physics it is laid down that one or more public lectures on subjects or branches of knowledge in respect of which study or research is being carried on in the School shall be provided for delivery in alternate years at University College Dublin and Trinity College Dublin. When this regulation was being drafted, A.W. Conway expressed the view that the preparation of a public lecture by a research scientist could be very time consuming. Fortunately Schrödinger and Heitler were not narrow specialists and so could draw on their reading to present, in an intelligible form, results of modern science.

Of the statutory lectures given in the early 1940s, those that made the greatest impression on the public were the series of four lectures given by Schrödinger at Trinity College Dublin in 1943 entitled 'What is life?' These dealt with the physical aspect of the living cell and especially with the bearing of the quantum theory on the structure of chromosomes and on the nature of mutation. The audience totalled nearly four hundred, and to accommodate them each lecture had to be repeated. The lectures were published and were subsequently translated into German, French, Swedish, Japanese, Italian and Russian. Some of the other lectures formed a basis for books published by Schrödinger and by Heitler on the relations between science, philosophy and humanism.

10 Burlington Road, site of the School of Theoretical Physics since 1971.

While the number of senior professors provided for in the establishment of the School was three, this number was not attained until 1948 with the appointment of J.L Synge. Already in 1945 Heitler had been raised to the rank of senior professor. The advent of Synge imported a more mathematical orientation to the research of the School and a massive development of research in relativity theory.

Further reading:
Institute for Advanced Studies Act 1940 (Dublin, 1940).
Institute for Advanced Studies (School of Theoretical Physics) Establishment Order 1940 (Dublin,1940).
J. McConnell, *Erwin Schrödinger (1887–1961). Austro-Irish Nobel Laureate* (Dublin,1988). Royal Dublin Society Occasional Papers in Irish Science and Technology no. 5.

Erwin Schrödinger, senior professor 1940–56.

ERWIN SCHRÖDINGER   1887–1961
Physicist

Born: Vienna, 12 August 1887

Died: Vienna, 4 January 1961

Family: Son of Rudolf and Georgine Emilia Brenda (née Bauer) Schrödinger
   Married: Annemarie Bertel 1920

Distinctions:
Membership of the following scientific
   academies:
   Vienna 1928, Prussian 1929, Royal Irish
   1931, Madrid 1935, Pontifical 1936, USSR
   1940, Lima 1944, Lincei 1947, Royal
   Society of London 1949
Honorary doctorates of University of Ghent
   1939, Dublin University 1940, National
   University of Ireland 1940
Medals: Medaglia Matteuci 1929, Nobel
   prize for physics 1933, Max Planck 1937

Addresses:
1940–1956   26 Kincora Road, Clontarf, Dublin
1956–1961   Pasteurgasse 4, Vienna

Erwin Schrödinger was a member of a cultured Viennese family. As a child he derived from his father an interest in botany, philosophy and painting and from his mother a proficiency in the English language, his maternal grandmother having been born at Leamington. His early formal education was chiefly in the ancient classics, and this helped him to become well acquainted with Greek philosophy. From 1906 to 1910 he studied at the University of Vienna, where he obtained an excellent training in theoretical and experimental physics from Fritz Hasenöhrl and Franz Exner.

Before coming to Dublin, Schrödinger had held University posts at Vienna, Jena, Stuttgart, Breslau, Zürich, Berlin and Graz. During his stay in Zürich (1921–27) he proposed what became known as the 'Schrödinger equation', which provided a means of applying the quantum theory of Max Planck to physics, chemistry and biology. In spite of his immense influence in spreading the knowledge of quantum theory, Schrödinger appears to have remained at heart a classical, that is pre-quantum, physicist.

Schrödinger's publications include sixteen books and about one hundred and sixty papers, many of which were translated into foreign languages. The range of his scientific publications embraces quantum theory, statistical mechanics, Brownian motion, dielectric theory, general relativity, optics. He wrote on interdisciplinary topics; in particular he investigated how physics and chemistry might be applied to biological problems. He was also very much concerned with the cultural value of the natural sciences.

Erwin and Annemarie Schrödinger became Irish citizens in 1948. Schrödinger left Ireland in 1956 to take up a personal chair in the University of Vienna.

Walter Heitler, senior professor 1945–49.

Addresses	
1941–1949	21 Seapark Road, Clontarf, Dublin
1949–1958	Drusbergstrasse 59, Zürich
1958–1981	Am Guggenberg 5, Zürich

WALTER HEITLER   1904–1981
Physicist

Born: Karlsruhe, 2 January 1904

Died: Zürich, 15 November 1981

Family:
Son of Adolf and Ottille (née Rudolf) Heitler
   Married: Kathleen Nicholson 1942
   Children: One son, Eric

Distinctions:
Membership of the following academies:
   Royal Irish 1943, Royal Society 1948,
   Leopoldina in Halle 1968, Mainz 1970,
   Norwegian 1974
Honorary doctorates of National University of
   Ireland 1954, University of Göttingen,
   University of Uppsala
Medals: Max Planck 1968, Marcel Benoist Prize
   1970, Literaturpreis der Stiftung für
   Abendländische Besinnung 1977, Gold
   Medal of Humboldt Gesellschaft 1979

Walter Heitler was born in Karlsruhe, Baden, Germany, of a Bohemian-Jewish family, nearly all of whom perished in the Nazi holocaust. His early education was classical but, at about the age of eleven, he began to develop a personal interest in the natural sciences. He studied at universities in Karlsruhe, Berlin and Munich, where he took his Ph.D. degree under the supervision of Herzfeld. After a brief stay in Copenhagen he arrived in Zürich just a few months before Schrödinger left for Berlin in 1927. Having mastered Schrödinger's papers on quantum mechanics he set about applying them to calculate the Van der Waals interaction between two atoms. He collaborated with another research worker Fritz London and the result was the Heitler-London theory of chemical bond.

In 1927 Heitler went to Göttingen as assistant to Max Born. When Hitler came to power in 1933, Heitler left Germany for Bristol where he remained until he transferred to Dublin in 1941. In the meantime he had been recognised as the world's leading authority on the quantum theory of radiation. In Dublin he devoted his energies to the theory of the newly discovered particle whose mass was about two hundred times that of electron—now called the muon. He gathered about himself an active group that included J. Hamilton, N. Hu, S.T. Ma, H.W. Peng, S.C. Power and P. Walsh. Though Schrödinger and Heitler were together in Zürich and Dublin, their research interests at any time did not coincide. Heitler became an Irish citizen in 1946 and retained Irish citizenship when he left for Zürich in 1949.

One of Heitler's regrets in his latter years was that he had chosen to specialise in physics rather than in biology or philosophy. Of his seven books, four deal with philosophy and religion, and he died a member of the Swiss Reformed Church. The number of his scientific papers exceeds eighty. It was a matter of surprise to his contemporaries that Heitler was not awarded the Nobel Prize.

James McConnell
School of Theoretical Physics
Dublin Institute of Advanced Studies